**OTHER BOOKS BY GEORGANN EUBANKS**

*Literary Trails of Eastern North Carolina*

*Literary Trails of the North Carolina Mountains*

*Literary Trails of the North Carolina Piedmont*

*The Month of Their Ripening:*

*North Carolina Heritage Foods through the Year*

*Saving the Wild South: The Fight for Native Plants*

*on the Brink of Extinction*

The Fabulous Ordinary

# The
# Fabulous
# Ordinary

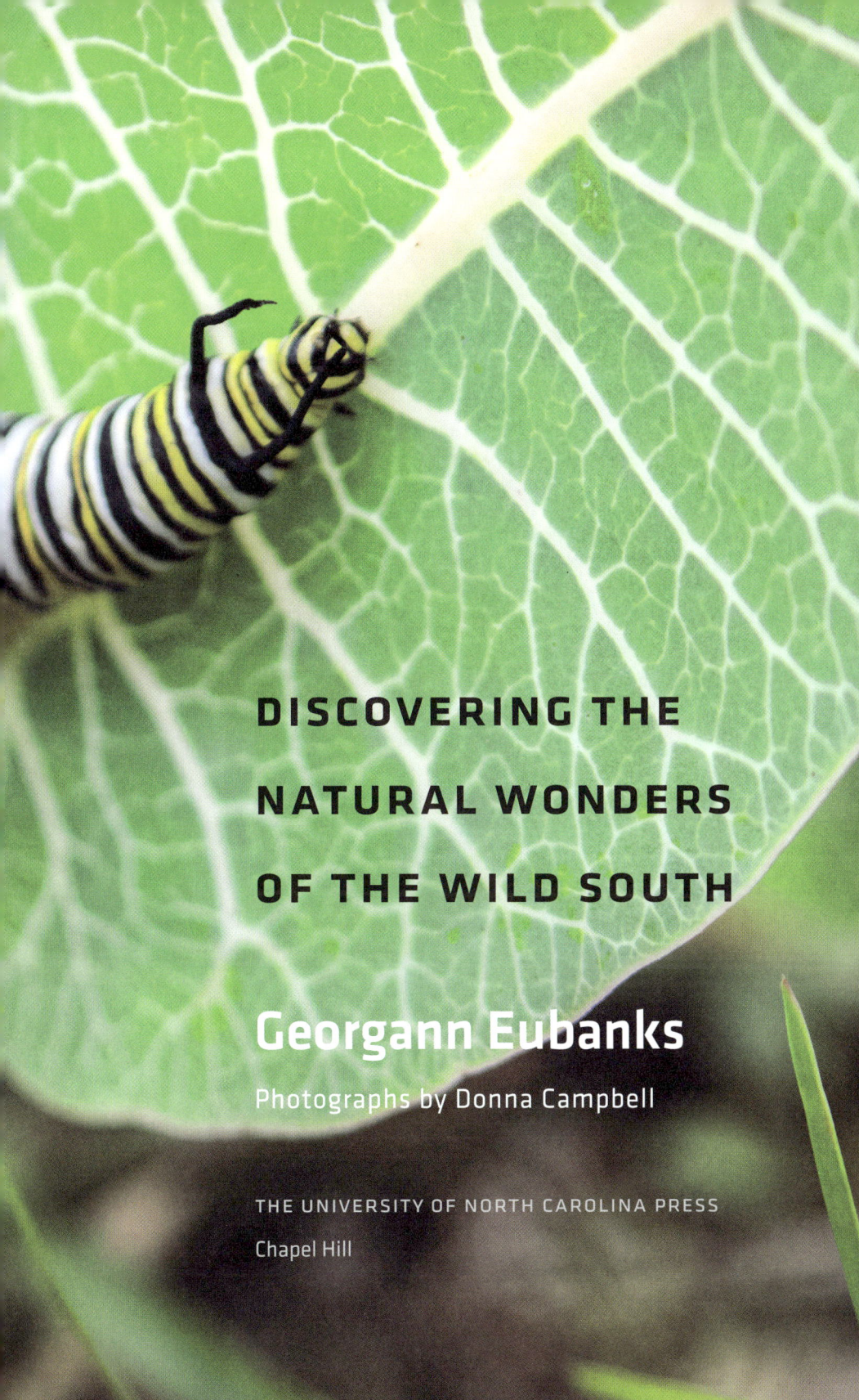

# DISCOVERING THE NATURAL WONDERS OF THE WILD SOUTH

## Georgann Eubanks

Photographs by Donna Campbell

THE UNIVERSITY OF NORTH CAROLINA PRESS

Chapel Hill

Set in Utopia and Klavika by Rebecca Evans
Manufactured in the United States of America

All photographs, except those otherwise noted, by Donna Campbell.

Ted Kooser, "Screech Owl" from *Delights & Shadows*.
Copyright © 2004 by Ted Kooser. Reprinted with the permission of
The Permissions Company, LLC on behalf of Copper Canyon Press,
coppercanyonpress.org.

Cover art: Purple martins and blackbirds arrive in huge numbers to roost
at sunset on Bomb Island in Lake Murray near Lexington, South Carolina.
Photograph by Donna Campbell. Used by permission.

Library of Congress Cataloging-in-Publication Data
Names: Eubanks, Georgann, author. | Campbell, Donna, 1951– illustrator.
Title: The fabulous ordinary : discovering the natural wonders of the
wild South / Georgann Eubanks ; photographs by Donna Campbell.
Description: Chapel Hill : The University of North Carolina Press, [2025] |
Includes bibliographical references.
Identifiers: LCCN 2024044990 | ISBN 9781469685922 (paperback) |
ISBN 9781469683829 (epub) | ISBN 9781469685939 (pdf)
Subjects: LCSH: Natural history—Southern States. | Natural areas—
Southern States. | BISAC: NATURE / Environmental Conservation &
Protection | TRAVEL / Special Interest / Literary
Classification: LCC QH104.5.S59 E93 2025 | DDC 508.75—dc23/eng/20241119
LC record available at https://lccn.loc.gov/2024044990

*For Wayne Stokes Goodall,*

*Susan Jane Larson, and Ellie Boudrow Perzel,*

*who always make the ordinary fabulous*

*What you look for in the world is not simply for what you want to know, but for more than you want to know, and more than you can know, better than you had wished for, and sometimes something draws you to a discovery, and there is no other happiness quite the same.*

—Eudora Welty, *Tell about Night Flowers*

# Contents

# Illustrations

# Preface

The fifteen connected essays in this book represent three years of travel and investigation. Our aim has been to document what are regular, ordinary events in the teeming ecosystems of the southeastern United States. The resulting narratives consider how plants, mammals, amphibians, and insects are managing to persevere despite immoderate pressures from human invasion, habitat destruction, and global warming. We also shine a light on the efforts of dedicated scientists, environmental activists, enthusiastic volunteers, and aspiring young naturalists who are working to help reverse the losses and preserve the fabulous ordinary that's still out there in the fields and forests, rivers and springs, wetlands and coastal estuaries.

The photographer, Donna Campbell, and I intentionally sought out common events that we had never experienced up close: the annual migration of a half-million purple martins roosting on a tiny island in a freshwater lake in South Carolina, the peak blooming time of thirty acres of dimpled trout lilies in a remote south Georgia forest, the few weeks in spring when gnat larvae glow like stars in the dark on the dripping rock walls of an obscure canyon in Alabama, and the overnight accumulation of dozens of elaborately patterned moth species on the side of a mountain cabin in North Carolina.

These and other wonders take place around us every year, and they are worth witnessing firsthand, especially amid the daily discouragement rendered in news stories about the fate of this planet that we have abused with abandon. We owe it to our children and grandchildren to offer these opportunities so that they, too, may experience the delights and surprises of this region.

After the worst global pandemic in more than a century, the urge to get outside and immerse ourselves in nature possessed many of us. COVID stalled air pollution, emptied office buildings, and kept us home to observe anew the nature around us. Our practices changed, at least for a while.

As we did our research, Donna and I also reflected on the insensitive culture in which we were raised—a culture that discounted and took for granted the expendability of the smallest living things around us. As a child, I ran around with neighborhood kids who thought nothing of stomping on an ant hill or pulling the logs off a cricket for no good reason. I took for granted that picking up shards of mica or smooth river stone for my rock collection was every kid's birthright. If we found a crayfish or salamander underneath a rock in the creek, we whooped and hollered and maybe poked at it with a stick.

I never thought much about pulling lichen or acorns off a tree to make a collage for grade school or jumping into the piles of leaves my father had raked up at the edge of the lawn. Scientists had not yet learned back then about mycorrhizal fungi and how it performs energetic recycling and soil-making in those leaves. My parents, from "the Greatest Generation," thought nothing of spraying the beginnings of every wasp's nest under the eaves of our house with the can printed with skull and crossbones that was kept under the kitchen sink. Now we have learned better, and we need to pass the lessons along.

In our last book, *Saving the Wild South*, we discussed the condition known as "plant blindness," in which many children and adults—immersed in their digital worlds—have never learned how to tell one green leaf from another. Developing the capacity to identify native species of trees, shrubs, and flowers was a resonant concept for readers. This guide adds another dimension by featuring dramatic birds, large and small mammals, amphibians, and insects that are native here or migrate through the region.

We admit that ecotourism, a growing enterprise in the South, does not always improve the environment for plants and animals. Nor can it instantly transform visitors into better stewards of the natural world. Some of the sensory adventures documented here reveal the clamor and damage we are causing as interlopers.

We must also acknowledge that individuals and families do not connect with nature in uniform ways. The privilege to travel at will and presume easy access to the natural world is by no means a universal condition. Race, class, gender, and prejudices have led to historic human conflicts that are steeped into certain sites and still shape the ways people approach the South. As a young woman, I was discouraged from heading into the backwoods alone. For people of color, some beaches, swimming holes, and parks—once segregated—harbor hard memories. Layers of grief buried deep in other sites—where violence has been perpetrated, particularly against Indigenous and African American people—must be recognized, even as new, artful gardens cover them.

Thick regional magazines promote tourism in the South and treat readers to romantic images and stories about how our temperate seasons, rich soils, and ample rainfall have created generations of avid gardeners, happy fishermen, and dedicated hunters, all celebrated at the appropriate season each year. Maybe attending to the avian guests whose migration routes now bring them here because of climate change—wood storks, roseate spoonbills, and sandhill cranes, among others—can move us to welcome and accommodate other newcomers who pass through or come to stay because of sea level rise and global warming. The landscape and the prospects of who can live where are changing dramatically for many species, including human beings.

The costs of our commercial culture of extraction, collection, and appropriation of the natural world are a big bill coming due. The accepted and often biblically justified superiority and separation of humans from all other creatures is being exposed as a dangerous philosophy. Our sense of entitlement to the supposedly endless riches of Nature has placed us on the edge of bankruptcy.

Here's how the radio producer John Biewen put it in his powerful essay "The Age of Dominion": "Humanity is rushing toward the ecological abyss. The question—the only question, really—is whether enough of us can mobilize, now, to force the dramatic societal change required to save ourselves. Whether we will demand the restoration of a culture that recognizes we are not the source—and that when it comes to nature's gifts, our rights and privileges are vastly outweighed by our responsibilities."

This book is a humble offering toward that end. Using these essays as a guide, and the photos as an enticement, we hope that parents and grandparents will follow our path throughout the US South and give the young explorers in their families the benefit of these experiences. The destinations and events described offer valuable science lessons that are in no way virtual. Groups of adult friends—young and old—who are looking for new and unusual trips with a certain measure of unpredictability and surprise will also find this book useful.

The destinations are spread across seven states: Alabama, Georgia, Florida, North Carolina, South Carolina, Tennessee, and Virginia. Some sites are relatively unknown but have interpretive services available to visitors. Some of the local guides we engaged were private entrepreneurs, and others were federal or state employees or nonprofit stewards working for the public benefit to protect fragile ecosystems.

Some chapters describe opportunities that are confined to a small window in a specific season and geography, but several chapters—those involving frogs, moths, butterflies, and owls—offer do-it-yourself experiences that can likely be set up without ever leaving home.

Every one of us can do something to repair the world, to use a phrase with deep roots in Judaic tradition. Every one of us, including future generations, deserves to experience the fabulous ordinary. An encounter with an elk—once lost to our region and now reintroduced—helps us see the past more vividly and understand the challenges of the present. Watching the sheer joy of once-endangered river otters at play reminds us how reasonable wildlife protections can still allow a species with commercial potential to flourish. We offer you these adventures to invigorate and inspire, to lead us all to a deeper reverence, and to move us to act on the vision of a planet that's hospitable to all living things.

# The
# Fabulous
# Ordinary

# 1

# Sandhill Cranes

*The world is awash with miracles, and I am thankful to simply bear witness.* —David Joy

We hear them already. It's a clattering sound, like a stick played over a washboard, a constant rattle underneath higher-pitched cries, calls, and honks. The sound of sandhill cranes feeding on the ground is like a traffic jam, but much more musical. We dodge briars and duck through winter-bare scrub to reach a small clearing, mulched with pine bark chips. This narrow stand of pines, oak, and gum trees is a buffer for the cultivated fields beyond. It has been a wet January, and the field is cropped to two-foot stubble. Water is puddled in gullies between the rows. The birds are jubilant.

We step closer, right up against a wooden wall. It's eight feet tall—a blind designed for photographers. Donna lifts a top-hinged hatch and latches it, resting her telephoto lens in the small, open window frame. Several others with cameras are already at work. We greet each other with nods and grins. I lift another hatch, a bit lower in the wall. The racket beyond these windows, no longer muffled, is palpable against my face.

Here in the 35,000-acre Wheeler National Wildlife Refuge, on the Tennessee River in Decatur, Alabama, an estimated 20,000 sandhill cranes have taken up residence for a few weeks. In early morning and late afternoon, they congregate in the fields to feed. It's a mixer, a mashup, a raucous dance party. Their roar and chatter drew us toward them the minute we stepped out of the car at the visitor center.

*Visiting sandhill cranes host a noisy January gathering in Alabama's Wheeler National Wildlife Refuge.*

The blind allows us to get very close—maybe twenty yards away. Some birds are moving even closer to us as the flock continues to expand at the edges. The birds bob and peck at the wet ground and at each other. Some pairs, I notice, stand beak to beak and mirror each other in a mime of movements. They remind me of hand puppets—beaks opening and closing in heated conversation. They jump up and down. Some raise and pump their wings as if the caller of this square dance has ordered "cross your arms and do-si-do," but there's hardly room for any pair to circle. They are jammed together in an ecstatic roil of bodies. They mostly move vertically. Threesomes preen to see which bird can stand taller. Only solo birds are beaks-to-the-ground, poking and pecking for nourishment.

Sandhill cranes (*Grus canadensis tabida*) are big. They stand three to five feet tall and have a wingspan of five to six feet. They weigh between 6.5 and 14 pounds. Cranes are one of the oldest living bird species in the world. Prehistoric cranes from 10 million years ago have been found in fossil beds in Nebraska.

As the afternoon light begins to change, more V formations glide in, a half-dozen birds at a time, whirling downward in spirals with barely a wing beat. The chatter is constant. Only when there's the sudden backfire of a car engine in the distance or the passing shadow of a possible predator—eagle or hawk—does the ruckus come to a dead stop. The cranes straighten their necks. They freeze and listen. For us, in the blind, the instant quiet is something like driving on the interstate through a pounding rainstorm and then rolling under an overpass for that moment of dry quiet before the rain pelts you again.

Then, just as suddenly, it's as if the DJ drops the needle and the crane party resumes. They crank it back up. The only other sound I notice is the whir of camera shutters on either side of me in the bird blind. This is a photographer's feast, a dream, and the afternoon light is just getting more flattering.

The cranes' feathers in this light are a steely bluish gray with short strokes of a rusty tan on their sides, as if watercolor had been applied with a brush. The wings, when they lift them, are much darker along the tips. These birds are easily distinguished from ibis or egrets, both in size and coloration. You could confuse them with a blue heron, I suppose, but adult sandhill cranes are bulkier and hold their heads straight out in flight, whereas herons tuck their necks into an S curve while standing still and in flight.

Sandhill cranes also have a long, straight bill and, most conspicuously, white cheeks and a spectacular red cap. That cap is not made of feathers but is red skin on the top of their heads. When the birds are excited, as they are at this afternoon mosh pit in Alabama, the red gets redder from the rush of blood to their heads.

During the nesting season, later in spring and summer, sandhill cranes go for a different look. They camouflage themselves by rubbing mud on their feathers, turning themselves to a reddish brown like the color of sand. Juveniles are also brownish before they mature. Unlike the heron, a crane's tail feathers curve down in the back to form what ornithologists wittily call a bustle, as in the padding that late nineteenth-century American and European women used to accentuate their rumps. Sandhill crane legs are long and black. For a photographer, the birds' coloration is fascinating because it seems to change in flight. Cranes appear more consistently brown on the underside when aloft, the brown often intensified by the muddy ponds they've visited.

These birds pair for life and nest on the ground. Their markings are mono-morphous, meaning both sexes look the same, though males are slightly larger than females, which was not an obvious distinction to my eye.

Sandhill cranes begin arriving here at the Wheeler Refuge in December and depart in late January for the Great Lakes and Canada, where they will breed and nest. The peak abundance of cranes at this site is generally in the first two weeks of January, though our visit on January 26, 2023, is spectacularly crowded.

Sandhill cranes travel in much bigger numbers on a Midwest flyway through Texas and up to the Platte River in Nebraska. There, come spring, the largest assembly of migrating cranes occurs—in the hundreds of thou-sands. Another flight corridor extends from northern Mexico and follows the Rio Grande in New Mexico. Cranes also make their way through Cali-fornia to Alaska and on into the Arctic.

The "greater" and "lesser" subspecies of sandhill cranes are distin-guished by their relative size and the distance they travel. The lesser cranes go farther north. They may be smaller in part because of the weight loss that accompanies such a rugged journey. The greater (larger) cranes tend to prefer stopping around the Great Lakes for their nesting habitat.

There is also a Florida sandhill crane (*Grus canadensis pratensis*) that stays put year-round. This nonmigratory subspecies numbers only be-tween 4,000 and 5,000 birds, according to recent estimates. The Florida cranes are legally protected from hunting in the state and were declared a threatened species in 1974 owing to ongoing habitat loss. The Florida cranes are joined annually by some 25,000 migrating cranes that visit the Sun-shine State for a time in winter but make their way up to the Great Lakes region from late winter to spring. In total, experts estimate that there are 750,000 sandhill cranes in North America.

The party goes on behind me as I walk back to the car to get a jacket. The temperature is dropping. Less than a quarter mile from the photo blind, a rushing stream of five o'clock commuters on the Point Mallard Parkway has now begun to add to the auditory chaos. Groups of migrant cranes are also flying over the roadway from the far side of the highway. They are assembling in a large group on the lawn in front of the visitor center and within sight of this raceway of vehicles. This open landscape is apparently irresistible to them. They seem oblivious of the traffic buildup, until some disturbance causes the group out front to take off all at once. The flapping of hundreds of six-foot-long wings makes a mighty sound like raucous applause, and every human coming and going from cars in the parking lot turns to watch them. The birds go up and circle, and then most drop down and resettle a few yards away from where they were to feed and loaf about.

We are only a half hour from the sprawl of Huntsville, a high-tech town that's expanding steadily south toward its airport, the beneficiary of renewed government spending on the Marshall Space Center and all the entrepreneurial start-ups that follow the defense and space industries.

A little more than 4,000 acres of the Wheeler Wildlife Refuge are within the official boundaries of the Redstone Arsenal. The arsenal, built in 1941, was once a chemical manufacturing facility and was later used for ballistic missile research and development. Now it is the daily destination for tens of thousands of government employees and contractors who work for NASA and the Departments of Defense and Justice. As we would soon discover throughout our travels for this book, some federal lands designated for wildlife conservation serve a dual purpose. The birds will take what they can get. This northeastern Alabama refuge is just east of the Mississippi flyway and has come to be an essential part of the sandhill crane's eastern migration corridor in the United States—a relatively new development in the life of these birds.

The Wheeler Refuge was set up in 1938 as a winter destination for ducks, geese, and other migrating birds. Wheeler Dam was positioned in the geographic center of the chain of Tennessee Valley Authority impoundments on the 652-mile-long Tennessee River, many of them TVA electrification projects from the New Deal era. The river begins at the confluence of the French Broad and Holston rivers in Knoxville; heads southwest toward Chattanooga; dips into Alabama; crosses the whole state; and then turns nearly straight up through western Tennessee into Kentucky to join the Ohio River near Paducah.

Wheeler was the first refuge created on a TVA reservoir that was "built primarily for electric power production, navigation, national defense and flood control," according to a 1949 pamphlet produced by the US Fish and Wildlife Service. "Most of the natural habitat of ducks, geese, and shore birds was being ruined by rapidly developing agriculture and industry," the document says. "Perhaps the new power reservoirs would in some measure create new waterfowl habitat to replace what had been lost."

Skeptics at the time doubted that these man-made wetlands would be suitable for wildlife, because of the need to change water levels periodically to produce electricity, and because the waters would be "subjected to intensive treatment to control malaria-bearing mosquitoes," the pamphlet said. US Fish and Wildlife Service biologists were brought in to consult, and malarial treatment practices were modified.

Still, it would take a decade for waterfowl populations to occupy the refuge in large numbers. Local farmers were key to the effort. They accepted the commission to plant several thousand acres of corn, milo, and millet in spring and winter wheat in fall. The farmers also agreed to leave part of their crops in the field for the winged guests. The refuge staff then began a practice of banding the wood ducks that come here each year, tracking the population to determine appropriate duck hunting limits in this part of Alabama.

The original size of the refuge has been reduced by about 10,000 acres since its founding. It sits to the east of the former shipbuilding city of Decatur and runs along a fifteen-mile corridor where the river serves as the boundary between the booming Alabama counties of Madison and Morgan. Hunting and fishing are popular here and are carefully monitored by the refuge managers.

Cranes did not start appearing until the early 1990s, according to Tom Ress, a refuge volunteer and local historian. The number of cranes was small at first but has steadily increased over the past thirty years. In 2023, the population of sandhill cranes peaked at 19,425 in January, Ress told me.

Breeding studies have shown that sandhill cranes usually lay two eggs each year, and the eggs incubate for about a month. Only one colt, as the chicks are called, is likely to survive. The male cranes are called roans and females mares. According to the International Crane Foundation, the horse analogy may have come from the observation that cranes, when they are not flying, run like horses, in a kind of gallop.

Crane families travel together until the juveniles are nearly a year old. Families and nonbreeding cranes form huge flocks as they travel north toward their breeding grounds in spring and back to their feeding and loafing grounds in winter.

Wheeler's mixed habitat, with its generous fields for feeding and mudflats for loafing and night roosting, clearly suits the sandhill cranes. The species does not roost in trees, and they like to have their feet in shallow water at night. They prefer sandbars in the middle of a river channel, where they cannot be ambushed easily by predators, which may include golden eagles, dogs, and large owls. Coyotes, foxes, raccoons, crows, and bald eagles are threats to eggs and juvenile cranes. According to Cornell Ornithology's "All About Birds" website, "Cranes attack aerial predators by leaping into the air and kicking their feet forward. They threaten terrestrial predators by spreading their wings and hissing, eventually resorting to kicking." Kickboxing cranes! I can imagine that scene based on the pugilistic displays in Wheeler.

Marshes, bogs, and wetlands where plants grow out of the water are crane favorites for feeding. The term for a group of cranes is a "sedge," which refers to the marshy habitat they prefer. Herons and roseate spoonbills fish for their food. In contrast, sandhill cranes eat seeds, grain, berries, tubers, insects, earthworms, and small vertebrates; they use their bills to probe for these morsels. (Florida cranes eat a "Florida diet," which may consist of nestling birds, snakes, lizards, frogs, and crayfish in addition to plant nourishment.)

The mixed diet of migratory sandhill cranes helps the birds increase their fat reserves for the long flight toward their breeding habitats. They use thermal updrafts to lessen their labors. Cranes fly at altitudes of 6,000 to 7,000 feet but can rise even higher—13,000 feet—when they cross the Rockies. They can fly forty-five to fifty miles an hour, and they migrate almost exclusively in the daytime, usually covering 200 to 400 miles in a day. At such heights and speeds, they need vocal power to communicate. Two friends who live in the crane flyways—one in Tennessee and the other in Florida—confirmed that *indoors* they can hear cranes calling to each other as the birds fly overhead.

At the blind, one of the photographers pulls back from the wall to talk to the group. He is excited. He has spotted a white bird on the far side of the field, at the edge of a shallow rivulet. He says the white one is likely a whooping crane (*Grus americana*), but he is waiting for it to stand tall and move a bit closer. We all turn to the wall to peer through our windows. He guides us to train our eyes on a slender tree leaning to the left at the edge of a stand of young oaks and gums, maybe fifty yards out. He says to let our eyes drop from the leaning tree to about eight o'clock. I see it. Donna sees it. The bird is slightly taller than the sandhill cranes around, but it keeps ducking its head, feeding in the knee-high stubble.

"Yep, it's a whooper," says another guy in the blind. With his giant telephoto lens, he can see the telltale bands on the leg of the bird. The whooping cranes, I later learn, are carefully tagged and monitored and have been for years.

"Wow," Donna says, angling to get a clear focus with her telephoto.

"We usually see a couple out here every year, mixed in with the sandhill cranes," says the veteran photographer.

I had read that whooping cranes, a critically endangered species, will sometimes travel with a sandhill flock. There is greater safety in numbers, especially when your bright white feathers make you stand out.

Then it hits me. The current estimate suggests that fewer than 600 whooping cranes are left in the wild in North America, and of these, some 73 are part of the eastern migratory population. "Fifteen of these individuals are wild-hatched and the rest are captive reared," according to 2024 estimates from the International Crane Foundation. I stare at the white bird in the distance. My eyes heat up and my vision blurs. I shake my head, then raise my shoulders in a shudder. What are the chances of this happening? I don't keep a life list like some birders, but this is a once-in-a-lifetime moment.

Then, just as improbably, a memory from elementary school overtakes me. It was a story we read about how whooping cranes were going extinct. The name was funny to my classmates. "Whoop, whoop," kids called out around the classroom and giggled. I was stuck on the word extinct. It sounded so severe, so abrupt.

There were enough times in elementary school when I was scared, sad, or helpless, but the whooping crane story we read that day felt like a loss that was unthinkable. How could a whole species go away? It was a big white bird I'd probably never get to see, not even at the newly renovated Atlanta Zoo or anywhere else, I thought.

I stared hard at the photo of the big bird in the newsprint story on my desk. Then I got mad. Why didn't we get to see those birds in real life? One day a helicopter had landed on the playground at my school, and these guys from a TV series hopped out to say hello. That was amazing. But why not these birds? Why couldn't they fly in or out? This deep if vague memory was now a sixty-year throughline from my childhood to this moment in Alabama. I am overwhelmed and grateful. I zoom in with my iPhone to try and get my own shot of the white bird. It has a red cap. Do whooping cranes have red caps, too? They do!

The story must have been in *My Weekly Reader*—those little newspapers rolled up in brown paper that the teacher would bring back from the mail room in the school office. She would tear off the wrapper and then smooth the papers flat before she passed them out, one per student. We'd usually be assigned to read something while my fourth-grade teacher left the room; she was a smoker. When she returned from the teachers' lounge, we'd have a lesson and discussion about what we had read.

That evening I looked it up. I was in fourth grade in 1964. Back then there were only forty-two known whooping cranes in the wild. Forty-two. That

*A rare and endangered whooping crane takes refuge among the sandhill cranes migrating through Alabama.*

same year, a young whooping crane was found on the Canadian tundra with an injured wing. He was rescued by biologists who were doing an aerial survey of the Wood Buffalo Provincial Park in Saskatchewan and Northwest Territories of Canada—still a critical nesting habitat today. Though the whooper's parents rushed the helicopter as it landed, trying to fend off the humans and protect their injured chick, the young bird would never be able to migrate south with his parents that season. He could not fly. Instead, he ran away through the marsh and was eventually captured by the wildlife surveyors.

His injuries were successfully treated, but he was still unable to fly. He would be forced to adapt to captivity and was taken to a wildlife facility in Colorado. His keepers named him Canus—combining the abbreviations of Canada and the United States.

In 1966, when I was in the sixth grade, Canus became part of a new endangered species propagation program at the Patuxent Wildlife Research Center, in Maryland. Unable to breed naturally because of his injuries, Canus became a sperm donor. US Fish and Wildlife biologists were then practicing techniques for artificial insemination, incubation, chick rearing, and proper nutrition. They hoped to build up a flock of whooping cranes in captivity as a backstop in case disaster came to the remaining birds in the wild.

By 1976, the year I graduated from college, there were only sixty-nine whooping cranes in the wild, but propagation science and conservation efforts were improving. The Patuxent breeding flock was growing, both by the incubation of eggs taken from Wood Buffalo Provincial Park and by the artificial insemination of captive females. Canus contributed to many insemination procedures in Maryland and at other facilities that received shipments of his cryogenically preserved sperm.

Sire Canus lived thirty-nine years in captivity, and his genetic material was ultimately involved in the births of 186 whooping cranes across several generations—children, grandchildren, and great grandchildren. At the time of Canus's death, in 2003, there were 420 whooping cranes in the wild and in captivity across the continent, ten times the number at the time of his capture.

As I soon would learn, the whooping crane we saw at Wheeler Refuge was part of the eastern migratory flock, an experimental population introduced to the wild from birds reared in captivity using the breeding techniques learned in the study of Canus. These birds were trained to migrate south by following ultralight aircraft to a national wildlife refuge in Florida. This remarkable story was documented in a film called *Journey of the Whooping Crane*, produced by Rhett Turner and Greg Pope for Georgia Public Television.

Welcome to twenty-first-century conservation. The reintroduction of these birds began in 2001, with captive-bred chicks raised by humans who wore white costumes and used crane-like puppets to imprint the fledglings, teaching them how to act like whooping cranes. The fledglings' migration escort, conducted by an ultralight pilot also dressed in a whooping crane suit, helped scores of cranes make their first journeys south. That improbable program came to an end with funding cuts, but a good number of whooping cranes raised in captivity are still following the migration cycle. According to the International Crane Foundation, "This population of around eighty wild whooping cranes breeds in Wisconsin and winters to the south, including Jasper-Pulaski and Goose Pond Wildlife Areas in Indiana, Hiwassee Wildlife Refuge in Tennessee, and Wheeler National Wildlife Refuge in Alabama." Reading that, I shivered.

*My Weekly Reader* ceased publication in 2012, but the whooper we saw would have made a great story for contemporary children. An internet search turned up several other folks around the country who remembered reading about whooping cranes in the same grade-school periodical over the years. The crane stories—which ran from time to time, tracking the plight of whooping cranes over two or three decades—made a lasting impression on many children. I read a blog by a science teacher who described the influence that the crane story had on her as a child, ultimately sending her into a STEM (science, technology, engineering, math) career in the classroom. An Alabama man named Chester McConnell launched and became the president of the nonprofit Friends of the Wild Whoopers, an organization that supports the whooping cranes in the Canadian park where Canus was born. McConnell read about whooping cranes in *My Weekly Reader* in the 1950s.

According to Tom Ress, the Wheeler Refuge historian and volunteer, whooping cranes have been showing up at Wheeler since 2003 or 2004. He said there have been as many as twenty individual whoopers that have visited over the migration period in recent years. A volunteer goes out weekly to monitor the whooping cranes and check their identities. All of them have been banded, so close observation allows the monitor to see the identification number of each bird. Some also have radio telemetry devices that were used starting in 2001 to track migration across the entire route, but the batteries in most of these devices died and have not been replaced. Monitoring must be conducted nowadays by visual sightings.

The next morning, we are back at Wheeler Refuge by eight. I am still in awe of this place. Near the fishing pier, a few dozen sandhill cranes are spaced out on a sand bar, probably where they spent the night. They are now lit by a warm morning light coming in low from the east. Their tall bodies are perfectly reflected in the calm water. A congregation of ibis decorates the limbs of a white oak leaning over the far shoreline, and a great blue heron is hunting in the shallows below them. The morning feeding ritual has begun, as has the river of traffic on the Point Mallard Parkway.

The sandhill crane is sometimes called "the ribeye of the sky" and touted as the best tasting wild bird available to hunters. In Alabama, where hunting is a long-lived ritual, hunting cranes is relatively new and only allowed in the northern part of the state—and only by Alabama residents, no outsiders. The bag limit per hunter is three birds per season, and the penalties for going over it are strict. There's an annual, computerized drawing for a limited number of permits and tags. Hunters must pass a test to demonstrate that they can identify the birds. They can shoot only in approved counties and not in state or federal refuges. Hunts can be conducted only in daylight hours and in two short windows on the calendar. Hunters are warned not to shoot into groups or at silhouetted birds with the sun behind them because of the possibility of killing a whooping crane. Tennessee has similar restrictions. Most sandhill crane hunting takes place in the western United States where the rules are not as strict and the sandhill cranes are much more plentiful.

We cross the highway from the fishing pier to the visitor center. Donna wants to go back to the photo blind, and I decide to walk to the Wildlife Observation Building, a relatively recent addition to the refuge. (The visitor center is closed for renovation.)

A gravel walkway leads from the parking lot into the woods beside the photo blind. The sandhill cranes are still making that rapid-fire clattering sound that is both percussive and musical. Flying overhead, they perform a call-and-response routine, which is two honks from the female to every one honk from the male, I read. It's how pairs stay in touch as they are navigating the sky. It's as if they are calling "Marco! Polo!"—a form of tag we used to play at the public swimming pool when I was a kid.

Sandhill cranes have some twenty types of vocalizations, and as I am walking, I'm trying to figure out how to name the different sounds. I even hear a purring sound, which whooping cranes also make, though whoopers' louder bugling calls are said to be different from the calls of sandhills. Now that I am only listening rather than watching the birds through the blind, I hear them differently, and the chorus does have a chanting quality that's hypnotic. It's both sacred and profane in its boisterous quality.

The trail to the building runs through a bog with a mature stand of river cane on one side and a low-lying wetland filled with vines, briers, and gum trees on the other. In the bog, a tree snag is being scouted by downy wood-peckers, while chickadees fuss and wrens and juncos swish about in the underbrush. I hear a blue jay and a towhee and then a distant train that must be passing through Decatur. I stop to watch the telltale wing flicks of a tiny ruby-crowned kinglet browsing for breakfast.

Beyond this tableau, there's a channel where mallards and coots are parading by in deeper water. I can see them through breaks in the scrub. I can also see the morning dance party of cranes on the opposite side of the trail through breaks in the river cane. Again, the white bird is among them. The same thrill as yesterday fills me. I bob and weave like a crane to catch glimpses of the bird through the canebrake. After a while a ranger passes by. I follow him. It turns out he is unlocking the observation facility. A sign at the entrance says, "Quiet Please." The building is a two-story box, set on a promontory overlooking water in two directions. Inside are tinted windows, carpet, and skylights.

From where I stand, I see sandhill cranes in a field beyond the chan-nel of water. Some are hanging close along the shore in groups. There are Canada geese, several species of ducks, cranes, and herons flying at inter-vals overhead, and they are loud. I can't help but notice that stereo speakers are mounted on the walls in every room. I wonder if this is a prerecorded tape loop of birds, but it seems random.

I find the ranger in a west-facing room outfitted with aluminum bleach-ers and a floor-to-ceiling wall of glass. It is amazing—an inside sanctu-ary for watching the outside sanctuary. The ranger's name is Brian Walch. He and his wife, Debbie, are full-time volunteers at the refuge. (Staffing, Walch explains, is hard post-COVID, as it is most everywhere, so the Fish and Wildlife Service depends upon retirees here.) They live in an RV on the property and perform light maintenance duties and engage with visi-tors. Brian and Debbie also have a house in Florida, where they spend a few weeks each year. They have worked at other refuges and are originally from North Carolina.

I ask about the wall speakers, and Brian explains that the building is out-fitted with exterior microphones, so what we are hearing is being piped in live. It is amazing. He offers some tips on where to see more ducks in the area, if I'm interested. They number about 50,000 in the refuge this time of year. Walch then excuses himself to get to his post in a tent in the parking lot outside the visitor center. There he is available to answer questions and hand out maps. His wife is going to lead a group of school kids on a tour of the refuge today, he says.

I stand alone in front of the wall of windows. No other human visitors are here at this hour. A threesome of cranes is no more than ten yards away from me outside. Two of the cranes are heads-down, feeding at the edge of the water, and the third one is standing sentinel. I am sure it can see me behind the glass and gives me a direct look. I note that its red cap is not bright but more like a dull, dried-blood color. It will not turn its back on me. When I move, it follows me with its head. Green-headed mallards and exotic-patterned wood ducks swim by. When a formation of cranes is nearing, I can hear them on the speakers and have time enough to raise my phone and capture the flyover with the camera.

I go upstairs to get a different angle, and I take some photos of the three-some from above. The sentinel is still looking for me.

This is not the wild. What is the wild anymore? Humans in a cozy building observing nature. That's what we've come to. Here, at least, the sandhill cranes and other waterfowl have a generous, human-engineered habitat, too. Loss of habitat is now the biggest threat of all, a loss also human engineered.

# 2

# Bald Eagles

LAKE GUNTERSVILLE, ALABAMA

*If facts are seeds that later produce knowledge and
wisdom, then the emotions and impressions of the
senses are the fertile soil in which the seeds must grow.
The years of childhood are the time to prepare the soil.*
—Rachel Carson

 Seeing a bald eagle up close—its size and dignity,
the sharp gaze of its yellow eyes, the strength
rippling through its wings—is unforgettable.
My impulse, however, is to avoid visiting raptor rehabilitation facilities and birds of prey
demonstrations. Animals in captivity depress
me. Injured raptors living in confinement is
usually a story that involves human interference.
The tales are hard to hear. At the same time, I
appreciate Rachel Carson's wish for children to
enhance their sense of wonder and to understand the increasing complexity of human relationships with other animals. It seems important to make
the effort to see a bald eagle living its life in nature to balance the vision of
injured raptors in captivity, the latter being the more common opportunity
available at wildlife centers.

Beyond the edges of the greater Huntsville/Decatur sprawl, we are now
headed farther south and east where the Alabama scenery becomes pastoral. Our destination, Lake Guntersville, is a little more than an hour away.
It's the state's largest lake (69,000 acres with 900 miles of shoreline) and
another popular resource for boaters and fishermen. Largemouth bass,
black crappie, bluegill, striped bass, and common bream are stocked here,

and fish-eating birds—especially osprey and bald eagles—hunt these waters. But it hasn't always been so.

As of the 1950s, fewer than 1,000 bald eagles (*Haliaeetus leucocephalus*) were left in the lower forty-eight states. State by state, a range of conservation efforts over the years helped to save the iconic North American bird, and the resulting species comeback has been impressive. Score one for the humans. According to a recent census I saw, some 200 bald eagle pairs would likely be in Alabama this January, their prime nesting season, though a sighting is hardly guaranteed, given the size of the area and the number of known nests—only three—in Lake Guntersville State Park in 2023.

Along the route, we pass modest farms, clapboard houses, and the occasional brick rancher. Most of the driveways are gravel. Functioning gas stations are few. Only the dollar stores look new. In my mind, I am still seeing the white birds at Wheeler Wildlife Refuge—those tall whooping cranes and the bright ibis posing on the branches of a white oak, perfectly reflected in still, sunlit water.

By early afternoon, it is growing cloudy, windy, and colder. White shapes along the roadside catch my eye. They are settling in the ditches and teetering in the fencerow scrub. I want to believe they are birds, poised to take off, but they are takeout boxes, wings unlatched, sprung open, and flapping in the breeze as if ready for a quick escape.

Did COVID lead us to litter more, I wonder? For months during the pandemic, on walks around my neighborhood, in Carrboro, North Carolina, I observed abandoned masks—those accordioned rectangles, yellow or light blue, the ear loops splayed like forlorn bird legs, soggy with rain. Seeing the face coverings was a reminder of the social divide between the cautious and the contentious. I realized then that I have always felt like litter is a form of social aggression.

More than three years after the peak of hospitalizations from the pandemic, takeout boxes seem to be the trend in roadside refuse. Restaurants everywhere have adopted grab-and-go as a regular option, even at fancy places that once offered only inside dining. Eat the meal, ditch the plate.

Where I live, in a green-leaning college town, the clamshells tend not to be squeaky Styrofoam but compostable paper. Whatever they are made of, they are multiplying on heavily traveled roadsides across the South, marooned in the bushes long after the food has been digested. I also notice in hotel rooms that paper cups are now individually wrapped, even at the free breakfast buffet. They are a chore to liberate, and they make for more plastic wrap to discard.

The claims made about the compostable products are not so simple, either. Sorting what is compostable is complicated. As a story in the *Eater* newsletter put it: "Compostables in the waste stream could, in fact, bungle up the composting process, create more trash, and continue consumers' addiction to single-use items, detracting from the most environmentally beneficial practices: reducing and reusing."

According to a *New York Times* story out of Seattle, bald eagles have been contributing to the nation's litter problem, at least in the Northwest. Juvenile eagles just learning to hunt have found certain suburban landfills irresistible. They, too, go for easy takeout meals, feeding on animal carcasses and food waste in the dump. Sometimes they have poisoned themselves on toxic materials in the landfill. The birds' natural tendency is to swoop in, pick up something, take off with it, and then drop remnants of the catch on suburban lawns, to the owners' great dismay. Occasionally, homeowners have been hit in the head by falling refuse.

Benjamin Franklin was critical of bald eagles for this tendency to scavenge and steal food from other creatures—a behavior that led him to complain to his daughter that America should not have chosen the bald eagle for its national bird. Franklin believed the wild turkey was more respectable, though he did not make his opinion public in 1782, when the selection of the bald eagle as the national symbol was announced.

Despite its mighty presence on official seals of the United States and on the flag of Alabama and several other states, the bald eagle lost favor for a time in the twentieth century. Salmon fishermen and livestock breeders declared the raptor a menace. As a result, by the 1930s, bald eagles were being killed by bounty hunters all over the country. The slaughter was massive. In 1940, Congress intervened and made it illegal to kill bald eagles or harm their eggs. Nevertheless, with concurrent habitat loss and the introduction of the pesticide DDT in the 1950s, bald eagles were driven perilously close to extinction. (DDT, it was eventually discovered, caused many avian species to absorb less calcium, making their eggshells weak and causing nesting parents to inadvertently crush their own eggs.)

According to a Library of Congress publication, the National Audubon Society counted only 417 breeding pairs of bald eagles remaining in the lower forty-eight states in the 1960s. Not until 1973, however, was DDT outlawed and the list of endangered species, including the bald eagle, created.

In 1973, there were no bald eagles nesting in Alabama. More than a decade later, in 1985, four juvenile eagles were released in northeastern Jackson County, in the state's first attempt to restore resident eagles. That same year, Linda Reynolds, the first naturalist hired at Lake Guntersville State Park, hosted the inaugural Eagle Awareness Weekend, an event now offered annually over multiple weekends in January and February. In the South overall, the number of bald eagle pairs counted in a recent census ranged from a high of 1,500 in Florida and 1,100 in Virginia to some 400 resident pairs in Tennessee and 200 in Alabama.

As Lake Guntersville comes into view, Donna is reading the agenda for the thirty-seventh Eagle Awareness Weekend, hosted by the Lodge at Guntersville State Park. Though COVID threw a wrench in these proceedings for a couple of seasons, the event is still the longest running interpretive program sponsored by the Alabama State Park System.

"The first event tonight is a wine tasting and cider social sponsored by a local vineyard," Donna says. "Then the park naturalist talks about the schedule of events for the rest of the weekend." She pauses and looks at me suspiciously. "Did you know this? Tomorrow morning starts with early-bird coffee and an outing at 5:30 *a.m.*!" (Donna is not a morning person.)

"We can do it," I say. "That's our first chance to see eagles. I hope there's a bus that will take us to the site. With coffee."

We make our way through the historic port town of Guntersville, with its marinas and hydroelectric infrastructure that began with the dam construction on the Tennessee River, completed in 1939. The look is vintage TVA, both shabby and charming.

At the far edge of town, we take a bridge across Short Creek. The road winds along the creek toward the park entrance. I am glancing upward, into the pine tops, hoping to spot an eagle nest. They can be enormous, wedged into the sturdy forks of upward limbs, assembled of sticks and moss and sundry ribbons of reed and grass. Eagles build nests as high as 80 to 100 feet in the air, and the birds return to them faithfully each year. "Nests have been found that are the size of cars, the average weighing up to 2,000 pounds," the Library of Congress story said. The biggest nest ever found weighed 4,000 pounds.

How very American, I think. Eagles keep making their nests bigger, dragging in new furnishings each season, and sometimes hoarding odd objects they pick up — things like children's toys, I've read.

The curving road to the lodge takes us up and around through a dense pine forest. Drivers are cautioned to go slow and watch for wildlife. Deer are browsing on a golf course that flanks the road. No one is golfing. The greens are brown. It is January, after all.

The lodge sits at the peak of Taylor Mountain. The entrance is fronted with massive stone pillars outside and matching pillars and oversized stone fireplaces inside. The floors are cut stone. The high ceiling and exposed wooden beams date from a 2004 renovation, but the grandiose style is familiar—the Alabama version of a Canadian hunting lodge. The cavernous lobby echoes the great outdoors and suggests the grand exploits of hunters of the past. Teddy Roosevelt comes to mind, and sure enough, the gift shop has teddy bears for sale, so named for the conservationist president who once refused to shoot an old, injured bear on a hunting trip in 1902.

Natural light pours in from the ceiling and through the bank of windows at the far end of the lobby, overlooking the lake. The view is dizzying. The lodge sits more than 1,000 feet above the water on a steep scarp, and it turns out that most of the rooms that fan out on three levels along both sides of the lobby have the same airplane view. Art deco stained glass pendant lamps hang from the rafters. The furniture is leather and oversized. Next to the Alabama flag on one side of the fireplace is a life-sized chainsaw sculpture of a bear (perhaps also a nod to wildlife conservation). A dramatic oil portrait of a bald eagle hangs over the mantel. Largemouth bass trophies are mounted near the check-in desk.

Later that evening, the line for wine is long, and the flavors are peach, strawberry, and watermelon. We pass. I had no idea what to expect of this gathering, but more than 200 people have assembled in the ballroom to hear the orientation session. There are few children; most folks appear to be retired. We sit near the front. Our host, Indya Guthrie, is a lively woman in her twenties, just completing her first year as the park naturalist. She explains that every year many folks are repeat participants.

Indya jumps right into Bald Eagles 101. The eagles here in Alabama, she says, are smaller than those in the North, where it is much colder, and more body weight is helpful. Most birds here range between ten and fourteen pounds. Females tend to be 10 to 20 percent larger than the males, Indya says. Bald eagles stand between two and three feet tall, and their wingspan is a remarkable six to seven feet. Ninety percent of their diet is fish, but on Lake Guntersville they will also prey on coots—a black aquatic bird that belongs to the rail family. Coots are cute; they swim like ducks but are smaller. They have a distinguishing white bill that extends upward like a shield on the face between the eyes. "The coots around here get up under the docks and hide when the eagles are hunting on the lake," Indya says.

Bald eagles mate for life and can live for thirty years or more. They begin to breed after five years and will produce young for about twenty years. Indya pauses. She points a remote at the ceiling. The lights dim and a video begins on the screen over her head.

"Eagle courtship is amazing, and occasionally we get to see this happen," she says. "Eagles will flirt while flying, and then . . ."

Two eagles, their legs and talons fully extended, are locked together in free fall on the screen. Wings out, they tumble like dancers, over and under, over and under. They might be skating partners or a circus act, but they are dropping—vertically and fast. Yet it feels playful.

This behavior is called the cartwheel display. The birds plummet from 100 feet in the air and let go of each other just in time to avoid hitting the ground or the water below them. The audience is gasping. "Show it again," someone hollers from the back of the room. We watch the clip again and applause follows.

Indya explains that once they mate, eagle pairs will take turns fishing and incubating the two or three eggs that will arrive over a week's time each year. The eggs are double the size of a chicken's egg. "Eagle chicks, once hatched," Indya says, "will gain a pound a week until they are a foot tall at about three weeks, but they don't fly until summer."

Because of predators, most couples manage to have one surviving chick each year. The juveniles are dark and look like vultures. "Eventually, they reach the dirty mechanic stage," says Indya. "The white feathers begin to emerge, and the juveniles look like somebody with a white T-shirt on who just did an oil change." What a metaphor! She shows us a slide.

Mature eagles have some 7,000 feathers, and the plumage is illegal to possess, even if an individual feather is found by chance. Steep fines are enforced. Native Americans, however, may keep the feathers for religious ceremonies.

"Eagle bones have air pockets," Indya continues. "They are lightweight and easily damaged. Eagle bones are only 6 percent of their total weight, whereas 20 percent of our human weight comes from our bones. Eagles can fly at 35 miles per hour and dive at 100 miles per hour."

Next, she projects a slide of two bird silhouettes on the screen before us. "When you are trying to tell whether you are looking at an eagle or a vulture coasting up high," she says, "remember that eagle wings are not curved but solid—they look like a two-by-ten when they are gliding." Another Alabama metaphor! Check.

The gathering tomorrow for 5:30 a.m. coffee, it turns out, will not involve special transportation. We will drive down the mountain in our own cars to the first destination in hopes of sighting a bald eagle in the wild.

It is not exactly an environmentally friendly start, this long caravan of rumbling diesel pickups, SUVs, and more modest vehicles, taking the curves down the mountain slow, headlights raking across the landscape in pitch dark when animals are usually foraging unmolested. There's a glittering frost on the ground, and when we get to the bottom, swirling scarves of fog rise from the waters of Town Creek. We drive a few miles away from town. At the marina, it takes longer to arrange all the vehicles on the grassy hill above the paved parking lot than it took to reach this spot. Already, a group of photographers in camouflage outfits has set up tripods with two-foot lenses pointing west, up the creek. They are regulars here, someone says, not part of our group.

The first sound we hear as we descend the frosty grass and assemble along the lakeshore is gunshots over the ridge, echoing across the water. It's duck season, someone explains. It's thirty-two degrees, and the early light is a sweet pastel mix—pale pink with gray-blue clouds to the west, yellow light rising to the east. The sun is not quite above the horizon yet. Some folks traipse out on the dock, hugging themselves against the cold. Others are huddled in clusters on the asphalt, their breath rising over their heads in clouds. There must be sixty of us, though some have stayed in their cars, running the engine and heating system, willing to watch through their windshields. Some soon head back to their cars for more insulation —blankets, scarves, toboggans.

Indya positions herself in the middle of the gathering and looks out across the water with binoculars. An eagle, about the size of a speck of pepper, has landed in a hardwood tree, empty of leaves and about a football field away. Indya points. Even at this distance, with my binoculars, the eagle seems huge compared to the limb on which it rests. Its white head glows yellow in the dim light. Indya suggests this eagle may eventually decide to fly toward us and have a look around the water beside us. We wait.

A group of seagulls meanders above the cove. The eagle, which Indya says is a female, is preening, plucking at her wings. After fifteen minutes or so she flies, but away from us, not toward us. And just like that, the sun is up. The group begins to disperse because it is so very cold, the kind that soaks into you. A hot, homestyle breakfast is waiting.

Back at the lodge, Donna and I find the huge dining room downstairs. We are ready for the indulgence of authentic southern biscuits and homemade gravy. The gravy is so good; as my stepfather used to say, "If you put that biscuit on your head, your tongue would beat your brains out trying to get to it."

We discuss our options for the morning and decide to skip the next group session in the ballroom. We have found a website posted by a local eagle enthusiast. The blogger explains that the Guntersville eagles did not know about COVID, and they would do their thing, whether people were there to watch or not. During the pandemic, he wrote up directions to a park downtown where chances to spot an eagle are decent, he said. His advice was to park at the waterfront (there's a large parking area) and look for a gathering of photographers standing under a tall pine. A nest could be found somewhere on Sunset Drive. Encouraged, we finish breakfast and a final cup of coffee and head down the mountain.

Scores of meandering mallards and coots are swimming along the Guntersville waterfront at this peaceful spread of shoreline. It is grassy, flat, shaded by giant pines, and carpeted with pine straw. Donna walks in one direction and I go the other way, both of us gazing up into the trees. It takes a while, but finally a man with a camera comes walking toward me from farther down the shore. I ask him where the eagle nest is, and he says we are in the wrong place. He points. It's on the north side of Highway 69, which divides Guntersville in half.

Later I will study an aerial map. Town proper sits on a peninsula shaped something like a fish. The west side of the peninsula offers the Sunset Drive trail, a 3.7-mile paved walking path looking across Lake Guntersville and the Tennessee River. It turns out we need to be on the northern half of the trail.

This photographer gives the same advice. We need to look for a gathering of people with tripods and cameras. "It's the biggest pine up that way, with woodpecker holes all around the bark," he says. "It's at Civitan Park, on beyond the ballfields." I make notes, thank him, and then catch up with Donna, who has been photographing the busy ducks.

We go back to the car and drive north on Sunset Drive. On the way, an empty osprey nest at the top of a pole near the ballpark is encouraging, but there's no cluster of photographers anywhere along the route. At the very top of the peninsula, we turn around. Heading back south, we spot what may be the biggest tree on the waterside of the road, and yes, there's a nest. I pull over quickly and park. We jump out. But the nest is empty. Abandoned. No noise. No photographers. No one is around. A lone jogger passes by. Guntersville Lake is very wide here, and the sun is coming in and out of the clouds, sparkling on the water. There's a breeze and, farther out, a few whitecaps.

Of course, I want a photo of the nest, and Donna is already working the angles. It is impressive, dense with sticks built into a funnel of forking branches, shabby with limbs protruding toward the bottom—a mighty big baby basket. I take pictures with my phone to remember the height and size, the scale. I walk around and around the tree. Donna is shooting from all angles, too. I study the woodpecker's signature, which looks like a machine gun has bored bullets into the bark. Minutes pass. All is still quiet.

When Donna has shot from every viewpoint and the light has come and gone behind the clouds several times, we go back to the car, a bit defeated. We sit a minute. I can't find my sunglasses. They have fallen forward on the dashboard. I stretch in my seat to reach them, and leaning into the windshield, I see motion. An eagle is coming toward the nest, a shredded ribbon of something in its beak and trailing below its tail. "Oh, oh—look!" I whisper. We open the car doors quietly. The eagle has not gone to the nest yet. It sits prominently in a pine across Sunset Drive. We stand there with our mouths open. Then Donna starts taking pictures. The bird adjusts its talons on the limb several times and then lifts, taking a wingbeat or two. It lights on the lip of the nest, then disappears into it. Donna moves behind the car, hoping to get a higher view. I cross the road for the same reason. A white head appears in the nest, barely above the lip. There is some action going on. Who knows what. We are mesmerized. Giddy. "The light's not great," Donna says in a low voice. We are both looking through lenses now. My phone. Her telephoto. The white head comes up higher and disappears again.

Then comes a second eagle from the same direction: maybe the male. I call out to Donna. With a lot of flapping in a tight space, he lands on a limb beside the nest. There are so many pine cones attached to the smaller limbs of the tree that it's an obstacle course for a bird so big. The sunlight is shining and then not shining. Traffic has picked up. Some drivers are slowing down to try to see what we are looking at. We just can't believe it. The nest had seemed dead empty. There must not be chicks yet. Maybe eggs?

The dance goes on for a while. First one moves and then the other. One white head appears just above the edge of the nest and goes back down. Finally, the male rises up and goes back to the limb beside the lip of the nest. He sits watching the other eagle we cannot see. Then the other white head appears. I motion to Donna to come across the street where I keep backing up toward higher ground. She darts across and moves behind me and stops, pointing her camera up. There has been no time to set up a tripod. Suddenly the hidden eagle takes off, wings arched, rising completely above the nest, the other still sitting, watching his mate. The sun shines

*In late January, a pair of bald eagles take turns preparing their nest in a tall pine along the shore of Alabama's Lake Guntersville.*

like a spotlight on the two of them in that moment. They found their light. Now the second eagle flies back into the woods across the road. It is 10:45 in the morning—what feels like a very full day already.

"Did you get that?" I ask Donna.

"I think so," she says.

The traffic has slowed to a crawl. A woman in a battered Japanese sedan pulls partway over on the lakeside and stops across the road from us. Four kids are in the car, leaning out the windows to get our attention. "What is it?" the woman says. "What y'all see?"

I dash across the street to her open window and pull up the best image on my phone. The little girls in the front seat—one appears to have just lost a tooth, the other has perfect cornrows—are wide-eyed. They scoot across and lean into their mother to see the image. The boys in back are jockeying for position, their arms grabbing at the back of the front seat. They are all close in age.

"It's an eagle," I say. "A what?" says one boy, all elbows, around his mother's headrest. The kids shake their heads, smiling big.

"We drive by here all the time to go to ball practice and games at the park," the young mother says, pointing ahead. "I seen folks looking up at something right here before, but we never knew."

We all grin and nod our heads. I am pointing up and the kids are now craning their necks to see the nest. "It is a very big nest," I say. "The eagle's wings were six feet across." I throw out my arms.

A horn sounds down the line of traffic behind us. Somebody has lost patience with our stalling traffic. I smile, back away, and the woman pulls very slowly into the road and drives on. "Thank you," she hollers, her arm out the window. A white man in a white truck lays on his horn all the way coming toward us. Then passing us, he guns his engine. I'm glad the birds have already flown away. The children are now looking out the back window, waving as they round the curve out of sight.

Donna definitely got the shot. We are thrilled. Later that afternoon, sitting in the great hall with the other conference participants, she leans forward and asks a middle schooler, a girl we had spoken to at breakfast, how many eagles they saw on the field trip we missed today. The girl makes a face and fashions a zero with her thumb and forefinger. Just then, there's commotion at the side entrance to the ballroom. Lodge staff are rolling dog crates in on dollies. It's the dreaded raptor show. David Haggard, a ranger with Tennessee State Parks, and Valerie Castanza, of Raptor Ridge Wildlife Education Center, in Eclectic, Alabama, have brought some of the birds that have become education ambassadors for sessions like this one.

Valerie got involved with raptor care as a volunteer at the Montgomery Zoo and Mann Wildlife Learning Museum after her youngest son graduated from high school. "I didn't expect it, but the raptors just drew me in," she tells me when we speak later. "I made it my goal to work with the zoo's golden eagle, Aspen. We would take her outside to fly on a leash. I learned what I could."

When Valerie and her husband moved to Elmore County, she launched her nonprofit aviary. She has four owls, a feisty red-tailed hawk named Grizzly, and a vulture. She takes the birds on the road regularly to present educational programs. She also volunteers at the Auburn Raptor Center, which is associated with the vet school on the Auburn campus.

David Haggard worked with his first eagle in 1988. When he began his career, he tells the assembly, three out of five birds that came in for care had been shot by people who saw them as competition for animals that they hunted for food. Nowadays, eight out of ten have been hit by cars. Roadkill is particularly tempting to raptors in winter, when food is scarce.

David is well known at this event. He has been presenting for years and has hosted tours at Reelfoot Lake State Park, in northwestern Tennessee, where a large eagle population lives. David explains that he learned his love of wildlife from his grandfather and father. He hunted and fished with them nearly every weekend of his childhood and early on decided that he wanted to spend his life outdoors. At the beginning of his career, he learned to climb trees to band eagle chicks, and he made several trips to Alaska as a member of a banding team. Over the years, he has brought several eagles back with him to rehabilitate. Today, Alaska leads the nation in bald eagle population, with some 30,000 breeding pairs.

David, who is approaching retirement, is wearing his official khaki ranger uniform and a green tie. He goes to the back of the room where the dog crates have been stationed. He pulls on a heavy leather glove and brings out Shadow, a barred owl that seems healthy. Shadow even gives us a guarded hoot. The bird's intense stare is unrelenting. Its head swivels as David walks around the room, gently grasping its feet. Shadow is also secured with a chain. People rise from their seats to photograph the bird.

Next comes Cypress, a great horned owl that cannot fly. One wing was damaged and had to be clipped years ago. As David walks Cypress around the ballroom, the bird opens both wings. The lack of symmetry is painful to see, and Cypress seems more excitable than Shadow as the crowd observes her.

"Great horned owls can kill things larger than themselves," David says. "They love chickens because they are delicious and not very smart. These birds have even been known to kill a turkey, but they can't carry it off." Laughter follows.

David, ever the educator, gives us a pop quiz. "What is the fastest animal in the world?" Lots of guesses are called out around the room, and all are wrong. It's the peregrine falcon.

Delta, a red-tailed hawk, is next. It is the most familiar bird to me. A couple of hawks have been living in my Carrboro neighborhood for years. David says the survival rate of rehabilitated hawks released from captivity is not good. Data suggest that in the region, some 75 percent rarely make it past a year or two, often suffering reinjury because of their weakened state.

Valerie Castanza explains that the choices for injured birds generally are euthanasia, placement in a rehabilitation center, or life as a teaching animal—the fate of the ones we are seeing today, who have been glove trained. Older birds cannot always be trained to tolerate the show-and-tell, however. The Alabama Wildlife Center, in Oak Mountain State Park, receives some 2,000 calls per year from people who have found injured birds, she says. Two-thirds of that number are songbirds that have experienced vehicle strikes, window strikes, or chemical contamination, especially from discarded oil. Some have gotten tangled in fishing lines or stuck in the glue traps that people set out to catch vermin. Cat attacks are another major hazard, which often result in puncture wounds, infections, and crushed bones.

"The goal is release for these birds," she says, but it is not always possible. Right now, the Wildlife Center cannot accept new raptors or water birds because of the threat from avian flu. She also noted that birds develop relationships with their handlers and show signs of depression if there are staffing changes. This labor of love is hard, relentless work. Put so starkly, I can now appreciate the value of these demonstration birds.

David's last bird is Bandit, a seventeen-year-old eagle that flew into a power line and damaged its wing. Bandit is the embodiment of power and threat, and David holds the bird with both hands, wearing gloves that reach over his elbows. But as David explains, bald eagles look far fiercer than their voices sound. Bandit is not happy, it seems, and is emitting what amounts to a high-pitched whistle. According to the US Fish and Wildlife Service, Hollywood has been known to substitute the vocalizations of a hawk, which are far stronger, when showing footage of the bald eagle, the American symbol. Still, Bandit's beak, which reminds me of a carpet knife, is not something I would want to encounter up close.

As Bandit goes back to the animal crate, a last cry ripples through the room. The voice is pleading. There is a great flapping and then the sound of the cage closing.

"It's always a hard decision," Valerie said to me later. "You don't want the bird to be in pain or stress, and having to make the decision about an injured bird is tough." She says that Grizzly, her red-tailed hawk with only one wing, "talks to other redtails that come by the aviary. He lets them know it is his territory. But if I ever felt he was depressed or not content in his role in life. . . . It's like having a pet, you do what you need to do."

# 3

## Dimpled Trout Lilies

*Once you learn to read the land, I have no fear of what you will do to it, or with it. And I know many pleasant things it will do to you.* —Aldo Leopold

Far below the bare-limbed trees of late winter in the southwest corner of Georgia, the first dimpled trout lily of the season will emerge. It appears as nothing more than a tiny, cylindrical blade of green, knifing up through leaf litter and fallen sticks. As the last hard freezes of the season come and go, a base leaf will also unfurl on the ground, speckled like the skin of a trout: hence the name, trout lily. The stem will eventually reach up six inches or so, and a purplish flower-head forms, shaped like a closed umbrella.

With sunshine, six petals of the umbrella open out. They are bright yellow on the inside and flecked with red around the middle of the blossom. Six red anthers, like tiny, ridged tongues, hang from the center of the flower, inviting bees and beetles to visit. As the day goes on, the petals curve back, much like a tiger lily, but the flower head is smaller than a sweet gum ball and dainty as a sulphur butterfly. It closes at dusk.

This doll-sized plant is easily missed in the wild if only one or two specimens are nearby. But what if there were millions of these diminutive yellow flowers, all in one place? That is what we have come to southern Georgia to see.

The better-known trout lily, *Erythronium americanum,* is larger and less fussy about location than the dimpled trout lily. It ranges from Ontario to North Georgia and west to Missouri and Oklahoma. The nineteenth-century poet Emily Dickinson wrote of the spring ephemerals that grew in her Amherst, Massachusetts, woodlands. As a teenager, she created her own personal herbarium, including a trout lily specimen — pressed and mounted on heavy paper. It can still be seen at Harvard University.

Both varieties of the trout lily have also poetically been called the dog-tooth violet because the tiny bulb that lives underground at the end of its root is shaped like a miniature canine tooth. Adder's-tongue is another name, given because of the blossom's tongue-like anthers. Dickinson used that name in her poem, "We Like March."

We like March, his shoes are purple.
He is new and high;
Makes he mud for Dog and Peddler,
Makes he forests dry.
Knows the Adder's Tongue his coming
And begets her spot.
Stands the Sun so close and mighty
That our minds are hot.
News is he of all the others;
Bold it were to die
With the Blue Birds buccaneering
On his British sky.

By contrast, the dimpled trout lily (*Erythronium umbilicatum*), which we have come to see, was not even described in the botanical record until 1963. It is more particular about its habitat. Botanical research suggests that across the whole planet it grows only in the seven southern states included in this book, plus Kentucky, Maryland, and West Virginia. And within those states, it's usually found in mountain habitats where the soil is acidic. It also spreads along Piedmont riverbanks where there's reliable moisture.

But improbably, in southern Georgia, very near the Florida panhandle, the dimpled trout lily prefers a north-facing slope, dappled with shade and sun. Here is what is believed to be the most prolific accumulation of the smaller lily anywhere on earth. Such a large population flourishing in the Georgia coastal plain is well below the southern extent of this species' normal range. The mass was discovered in dense woods between a small creek and a state highway, which runs from the well-known city of Thomas-ville (population 21,000) to Cairo (pronounced KAY-ro, population 10,000) to Whigham (population 423). The lilies live between Cairo and Whigham.

It's the day before Valentine's Day 2023 when we meet the revered local ecologist Wilson Baker at the lily site along with several other local conservationists who worked for more than a decade to create the Wolf Creek Trout Lily Preserve, just off US 84. Baker, now officially retired from his long-running position at the private Tall Timbers Research Station in Tallahassee, has also served as a consultant for many years with The Nature Conservancy. His curly hair and neatly trimmed mustache may be snowy white, but he is still an agile outdoorsman.

"When they four-laned US 84 in the 1980s," Baker says, "a biologist with the Georgia DOT saw a plant he didn't recognize on the roadside." The state employee called Angus Gholson, a well-known botanist in the region. Gholson served as the resource manager of nearby Lake Seminole, an Army Corps of Engineers project that created a reservoir in 1958 at the confluence of the Chattahoochee, Flint, and Apalachicola rivers on the Florida state line. Gholson came over to the highway construction site, about forty-five minutes from his home, and found a small population of the dimpled trout lily and another spring beauty, the spotted trillium. The flowers were within sight of the new highway but on private property.

"It definitely got his attention. Angus knew it was a special distribution of these plants in an unlikely place," Wilson Baker says. It would be Baker himself who ventured farther back into the woods at bloom time one February some years later and discovered the amazing numbers of plants. By early 2006, Baker had brought Dan Miller, a retired chemist from Ohio who had established his own native plant nursery, Trillium Gardens, in Tallahassee, to the site. Miller was wowed by the lilies, normally found in the Appalachian Mountains.

"I'd been around trout lilies in North Georgia, and I'd seen what were considered large populations up there," Baker says. "But nothing like this." For his part, Dan Miller was also taken with the property's profusion of trillium, another spring ephemeral that Miller had been growing on his property in Tallahassee. He eventually counted the trout lilies in a few, small, like-sized squares to determine the average number of plants per square foot and then multiplied that number by the full area covered by the lilies —more than ten acres. He estimated the total number of lily plants here to be in the tens of millions.

Baker continues the story: "We got to talking to other experts, including Tom Patrick, who was then the chief botanist in the Georgia Department of Natural Resources. He thought like I did that the number of plants here was very unusual."

By this time, the owner of the 140-acre property, Flint River Timber

Company, was preparing to sell the acreage in ten-acre tracts for a housing development. Dan Miller initially proposed that they put together a group of citizens to work out a deal with the timber company to buy only forty acres—the sunny slope where most of the lilies had taken up residence. However, development of the rest of the tract, they realized, which was on higher, flatter ground, would probably cause erosion, and human use would cause further disturbance. The lilies would surely be diminished.

In January 2007, Flint River Timber selectively harvested some trees on the property and built logging roads into the tract. Dan Miller, Wilson Baker, and Gil Nelson, all members of the Magnolia Chapter of the Florida Native Plant Society in Tallahassee, put together a field trip the next month when the lilies would be at their peak so that others, including the landowner, could see what the full spread looked like. The timber company had never visited the property during the blooming season.

The lily advocates—at this point, Baker, Gholson, Miller, and Nelson, joined now by Beth Grant, a devoted naturalist who lived in Thomasville— began strategizing about how they might acquire the entire tract. Grant was working hard at the same time to preserve the nearby Lost Creek Forest in Thomasville, after Thomas County officials had announced that they were considering the development of an industrial park there. Lost Creek is an intact climax forest with hundred-year-old hardwoods at the headwaters of the Aucilla River. The land also has a healthy population of the endangered Florida milk vine (*Matelea floridana*), which is endemic to Florida and occurs in only a few counties in southern Georgia.

The Lost Creek Forest was successfully spared from development, and in 2008, Beth would become the founder and first president of the nonprofit that still oversees the property. Grant has served as a frequent hike and workshop leader there and at the nearby Birdsong Nature Center, a 565-acre site also dedicated to environmental stewardship, a diversity of habitats, and the native wildlife of the region.

Both the Lost Creek Forest and Birdsong projects involved much larger tracts of land than the Wolf Creek parcel, and the lily champions soon discovered that their project was too small to qualify for most state and national grant programs. Many local people pledged donations to the fund to save the lilies, but no single donor among the group could manage the whole purchase price of more than a half-million dollars.

Thanks to state botanist Tom Patrick, the group then discovered the Georgia Land Conservation Program, which worked at the county level. They wrote a proposal for a one-time grant that would provide half the funds required to buy the Wolf Creek property, provided they could raise the rest of the money. The property would then be deeded to Grady County to create a preserve.

County leaders agreed to the idea and submitted the grant proposal, but they cautioned the volunteers that local government would not invest further in the project other than changing the property's tax status to not-for-profit. The grant was approved in December 2007.

By the time the team had put all this together, the national economy had begun to decline, and local property values along with it. The timber company that owned the Wolf Creek land kindly agreed to reduce the price and sell the property for the original purchase price, taking no profit. Still, as the deadline approached to acquire the matching conservation grant for the purchase, the campaign was short $45,000.

"They were going to shut down the grant program. We were the last group to apply for that pool of funds," Beth Grant tells us that February day in 2023 when we met her and some of the other conservators at the site. "Several of us gave a thousand dollars apiece," she says, but it was yet a steep climb. Getting the local garden clubs and prominent families in the region on board took time.

Dan Miller, standing with us at the site that day, chuckles, remembering the many small gifts—$50 or $100—that dribbled in. He says he was always confident they would meet their fundraising goal, but time was running out.

One energetic volunteer, Betty Jinright, of Thomasville, who was a former president of the State Garden Club of Georgia, had given countless tours of the land during bloom time in those first years after the discovery of the wildflowers. One day, rifling through some papers on her desk, Betty came across a seed packet with a phone number that was scrawled on the outside. She called the number, hoping to jog her memory about the seeds. The phone number belonged to a friend in North Georgia, and during the call, Betty invited her to come see the lilies. The woman accepted and came for a tour. She was thrilled by the site and told Betty that she owned several thousand acres in North Georgia where the lilies also bloomed, but not in such profusion. As her friend was preparing to go back home, Betty handed her a fundraising solicitation letter, and that was that.

Weeks passed, and then Dan Miller received a phone call from Betty's friend, who still had the letter but had forgotten how much was needed to buy the land. Dan told her, and she thanked him and hung up.

Several days later, a check for $45,000 arrived at the Grady County administrative offices, enabling the group to secure the matching grant. The lilies were saved in the nick of time. To the scrappy volunteers who had worked so hard raising funds, the $45,000 seemed as miraculous as the fragile flowers that stretched in every direction from the trails through the property that they had begun mapping.

"We had a backup plan," Dan Miller confesses. "We were prepared to get a loan for the amount we still needed to secure the grant. But we might still be raising money today to pay off that loan with interest," he says, grinning.

The Wolf Creek Trout Lily Preserve opened to the public in February 2009.

Our trip to the preserve in 2023 was our second visit. In 2019, when Donna and I were doing research for our book *Saving the Wild South*, we stopped to experience the lilies on the way home from Florida. Helen Roth, a native plant expert and dedicated member of the Magnolia Chapter of the Florida Native Plant Society, had told us about them and encouraged us not to miss the Wolf Creek preserve.

Heading north from Tallahassee early in the morning, we called ahead to ask if we could get into the gated preserve on a Sunday. The volunteer who answered the phone, Margaret Tyson, told us it would be best to see the lilies after the morning sun had worked on the woodland tract and warmed the slope. But we had a long drive ahead of us that day, and Margaret was kind enough to agree to meet us there midmorning—literally to see what was up.

Margaret is the principal volunteer who answers calls generated by the preserve's website. She comes from a sturdy family that has lived in Cairo for generations; her house was built from handsome timber harvested in the area by her grandfather. A mental health counselor by trade, Margaret was also a standout player on the University of Georgia's first women's basketball team to operate under Title IX in the early 1970s. After a stint as a college coach and counselor, Margaret came back to Cairo and has been active with the Apalachicola River Keeper organization and other regional environmental projects. She continues to provide private mental health counseling and consultation services.

Our timing in 2019 was fortuitous, Margaret said. The lilies were moving toward peak numbers. It was the day after Valentine's. The sun was bright, and the flowers had just begun unfurling in the damp morning. I didn't know what to expect. They seemed so very small when I came near the first patch along the trail.

To see the full picture, we had to stop and peer beyond narrow pathways marked with sticks as guardrails so we wouldn't step on the emergent ephemerals. I did a double take. The lilies were spread out in every

*At the Wolf Creek Trout Lily Preserve near Whigham, Georgia, the massive blooming tends to peak around Valentine's Day. Photo by Margaret Tyson.*

direction, well beyond our human capacity to see them all at once. I'd swear the lilies were yawning and stretching open as we turned in circles, taking in a view that was like watching a time-lapse film in 3D. The mottled tree trunks—large and small—were the only interruptions in this glorious ground cover: a carpet of bright green, dotted with emerging strokes of yellow. It was like nothing I'd ever seen.

Margaret said that with a spread of plants as large as Wolf Creek preserve, the blooming period can last for a few weeks. Plants of different ages stagger the timing of the show, joining in along the way to create the effect of a painterly crescendo toward peak bloom time. Once the spring leaves fill in the trees above, however, the flowers and foliage melt away, leaving whitish, dimpled seed pods—"about the size of chickpeas," as Beth Grant described them to us on our second trip. "If you'd come two months later," Wilson Baker told us, "you would not have seen any sign of them. Even a good biologist would not know they're here."

Though the lilies do have the tiny dog-tooth bulbs underground, called corms, the seed pods seem to be the most vital part of propagation. After blooming season, the dimpled fruits are carried off by ants that eat the sweet coating and leave the seeds underground in their trash pile, which— shall we say delicately—includes ant-made fertilizer. This symbiotic partnership between ant and plant is known as myrmecochory, pronounced "mur-mee-CO-coree."

There's a catch. Each seed may take five years of germination before the first flower emerges. So, the dimpled trout lily is not a species for the impatient. To find it spreading in the wild on its own clock, however, is a joy.

Margaret called our attention to several other exquisite spring ephemerals on the land. The native spotted trillium (*Trillium maculatum*), also known as the spotted wake-robin, has three mottled leaves, set high on the stem. The leaves look as if they've been painted with army-style camouflage. A deep burgundy bloom stands up at its center. This plant grows in rich forests of the deep South and blooms in concert with the trout lilies. It was abundant on both our visits.

Elsewhere on the preserve, botanists have found the green fly orchid (*Epidendrum conopseum*, also known as *Epidendrum magnoliae*), a perennial herb that, remarkably, grows only on the trunks and limbs of magnolias and live oaks. It does not bloom until later in the spring—round about May. Margaret noted that the preserve is staying open a bit longer in the season now than in past years, providing visitors a chance to see the other spring wildflowers that come later in waves on the property, some down the slope and closer to the bottomland at the creek.

In late February and into March, two other diminutive spring ephemerals appear: the coralroot orchid (*Corallorhiza wisteriana*), which doesn't have chlorophyll and is, thus, reddish rather than green, and the thin, southern twayblade orchid (*Neottia bifolia*). They are hard to spot because they blend into the leaf litter.

The showy jack in the pulpit (*Arisaema triphyllum*) and the bloodroot (*Sanguinaria canadensis*), with its bright white flowers with yellow centers, come along in March. The latter continues blooming until May, as does one of my favorites, the rain lily (*Zephyranthes*), also known as the atamasco lily or rainflower. I remember being surprised and delighted in our travels by these blossoms, which seemed to burst forth willy-nilly on the roadside in rural Florida right after a good spring rain.

Carloads of visitors arrived all afternoon on the blustery February day of our second trip to the preserve. I wondered why so many plant enthusiasts were drawn to this sparsely populated land of rattlesnakes and coachwhip snakes, of tornadoes and hurricanes, where a hard living was barely made by subsistence farmers who settled here after the Indigenous people were banished.

The likely answer is because so many unusual species are here to study and admire. "Place is fate," as the writer Frances Mayes put it. She grew up in nearby Fitzgerald, Georgia, and is the author of the memoirs *Under the Tuscan Sun* and *A Place in the World*.

The territory along the Georgia/Florida border, sometimes called the Wiregrass region, also known as the Red Hills, is rolling and surprisingly steep in places. Some roads dive and rise again like a roller coaster. The soil composition is rare, too. Attapulgite clay—a magnesium-aluminum silicate—occurs only here, and it is strip-mined for its absorbent properties used in products ranging from jet fuel, automotive paint, and pharmaceuticals to olive oil, skin cream, and kitty litter. Here, too, the weather can be quite harsh; hurricanes and tornadoes create massive wind damage to trees, and floods erode the red clay hills.

Human history here has been shaped in response to the unusual characteristics of the land and weather. In the 1890s, vast tracts of longleaf pine grasslands with great commercial value were bought up by American industrialists who lived elsewhere. Other areas challenged human settlement—swampy wetlands with lime sinkholes and underground rivers. A bit farther south and west, on the banks of the Apalachicola River, sandy steepheads—described on the Florida State Parks website as "deep ravines that occur in upland sandhill or clayhill habitats"—host several endangered species of plants and trees that grow nowhere else on earth, but the steep banks are not easily navigated by humans.

In geologic time, this area is the youngest part of Georgia. According to the *New Georgia Encyclopedia*, the region was underwater one million years ago, "forming part of a sea whose shoreline corresponded with the present fall line. As the ocean receded, a series of five sandy terraces emerged above sea level. The highest of these, the Hazlehurst Terrace, reaches an altitude of about 215 to 260 feet." Plants that don't "belong" at these latitudes, such as the dimpled trout lily, are likely here because of the same glacial upheaval that created the Appalachian Mountains. "This terrain has allowed certain northern plants to find their range elevation and soil type here," Wilson Baker told us.

Highway 84, which flanks the Wolf Creek Trout Lily Preserve, has been designated the Georgia Wiregrass Parkway. Wiregrass (*Aristida stricta*) is a native round-bladed perennial that grows in clumps that bend outward and stand more than a foot tall. In these pinelands, wiregrass was a dominant groundcover plant. However, as Baker pointed out, you aren't likely to see the plant along this stretch of highway, because of many years of disturbances. "The wiregrass has been drastically harrowed," he said.

Vast tracts of longleaf pine forests have been carefully—some would say artistically—maintained here. They comprise the largest remnant of what was once a 90-million-acre longleaf ecosystem in the Southeast that stretched from Virginia to Florida and west to Texas. The biodiversity in this remnant region of the longleaf is a major asset, but it requires intelligent and innovative management. The application of regenerative fire, which the native peoples in the region probably used each year to foster new growth and to attract wildlife for hunting, is being practiced again on an annual or biannual basis, and selective timbering keeps the forests looking their best.

Walking through a longleaf pine savannah is a deeply sensory experience. Because of the heady fragrance of longleaf pine needles—not harsh, like pine-based cleaning products—they are sometimes used in cocktails and desserts and to make tea. Native people knew their medicinal value. They are known to lower blood pressure and soothe sore throats. They are antiseptic, antioxidant, diuretic, and high in vitamins A and C.

Among the earliest known inhabitants of the Wiregrass region are the Lower Creek and the Hitchiti peoples. The Lower Creek peoples used this area for hunting and fishing. The Hitchiti are said to have met the Spanish explorer Hernando de Soto in 1540 on the Ocmulgee River. According to the Peach State Archeological Society, in 1733 two delegates of the Hitchiti tribe "were noted as accompanying the Lower Creek chiefs to meet Governor James Oglethorpe at Savannah."

Private collectors have found remarkable artifacts from the Hitchiti, who lived in settlements along the east bank of the Chattahoochee River, which forms the border between what is now Alabama and Georgia at midstate.

The Wiregrass region was laced with railroad lines by the early 1800s, offering a means to harvest and remove the valuable longleaf pine that dominated the ecosystem here. Extracting the lumber, turpentine, tar, and pitch left huge barren swaths, and for a time before the Civil War much of that land was given over to raising cotton. Corn and cattle became the predominant crops after emancipation. Some seventy cotton plantations were functioning in the Thomasville area at the time of the Civil War, but by the 1890s, northerners had bought them and converted them to private hunting preserves and retreats.

Before the railroad ran down the East Coast all the way to Miami,

Thomasville was a winter resort destination. From this history it has developed unusual amenities for a town its size: high-end retail stores, a top-ranked medical system, and historic houses immaculately kept for the part-time tenancy of their owners. Thomasville is a destination for rose lovers, who have been coming in the thousands since 1922 for the Rose Show and Festival every April.

Although the wild quail population fell into decline in the 1940s, such was its earlier, legendary abundance that quail hunters show up here, too, from October to March. Research in the Red Hills area of Georgia and Florida by the conservationist and ornithologist Herbert L. Stoddard (1889–1970) showed how regular, controlled burning of pinelands is essential for successful quail habitat. Aldo Leopold, often identified as the father of wildlife ecology, was a mentor to Stoddard, who helped to establish wildlife management as a profession and was the cofounder of Tall Timbers Research Station in 1958. Tall Timbers hired Wilson Baker as its first full-time biologist in the mid-1960s.

Today, "plantation" is the word used to describe the pinelands—including longleaf and loblolly landholdings—around the region. "The early plantation owners believed in Smokey the Bear [considering forest fires a scourge] and wondered for a time where all their quail went," Baker told us. "Stoddard put it together." Now these properties are being managed with controlled burns and selective harvesting, keeping the understory open and aesthetically vibrant. The quail population is slowly increasing, as a result. The burning seems to have given the endangered Bachman's sparrow a second chance, too, after the bird nearly disappeared when the burning practices were suspended in earlier decades, Baker said.

Visitors coming to see the lilies will also find the finest example of a longleaf forest, which is along the route from Thomasville to Cairo. Greenwood Plantation—purchased in 2015 by Emily "Paddy" Vanderbilt Wade, a lifelong conservationist—protects some 4,000 acres of forest. Part of the purchase was a 1,200-acre tract called the Big Woods, comprising old-growth longleaf. Some trees here are between 400 and 500 years old. The Greenwood Research Foundation manages the land and supports research on the ecology of longleaf yellow pine forests.

You might also consider a visit to Birdsong Nature Center where visitors can hike the pine-scented paths, perhaps chew on a few young green needles to get a taste of the place, and learn more about that unusual ecosystem.

Wilson Baker, a native of Pennsylvania who studied at Earlham College in Indiana and at the University of Georgia, began his work at Tall

Timbers with bats and other small mammals. His career interests expanded over the years to encompass the entire longleaf ecosystem. He has long been involved in protecting the red cockaded woodpecker and conducting butterfly surveys. A local beetle, *Aphodius bakeri*, was named in his honor. In addition to his role with the management team in organizing and maintaining the Wolf Creek preserve, Baker has documented endangered plants in Georgia and Florida, and most recently he discovered in the region a new species of coreopsis, which also bears his name, Baker's tickseed (*Coreopsis bakeri*).

At Baker's suggestion, we drove over from Wolf Creek to see the town of Chattahoochee, Florida, where Angus Gholson's family home sits at a high point above the town. The sprawling white house, with its appealing porches and many bright windows, is a replica of the original where Gholson was raised, which burned when he was a boy. Damage from the fire is evident on the giant live oak in the side yard.

"Angus used to tell us how a horse and buggy with a water tank came up the night of the fire and poured a little water on the house, but it was no use," Baker said. "Longleaf pine is the finest wood for building. Its heavy rosin repels insects, and it is very inflammable."

Gholson, who died at the age of ninety-two in 2014, was revered as a storyteller. He was also an aficionado of southern cooking, especially desserts, and ice-cold Coca-Cola in six-ounce bottles. He and his wife, Eloise, who turned ninety-eight in 2023, would entertain guests outside on the porch. They welcomed strangers from around the world who were drawn there by the fame of Gholson's botanical knowledge of the Panhandle. The house is still maintained by the family, though Eloise moved to an assisted living facility.

From Gholson's house, on Bolivar Street, the narrow neighborhood lanes slope down precipitously and lead to the 126-acre Angus Gholson Nature Park, established in 2003 to honor the conservationist. Gholson played in these sandy woods as a boy, and he learned to identify wildflowers that were endemic to this habitat, such as the endangered fringed campion (*Silene catesbaei*), a lacy confection that looks something like a miniature pink doily in bloom.

A short trail runs from the park along the bluffs and ravines of the upper Apalachicola River, where trout lilies also bloom this time of year. Other trails in the system are being repaired by volunteers from the Magnolia Chapter of the Florida Native Plant Society following serious damage to the overstory by Hurricane Michael in 2018.

Driving back to the Wolf Creek preserve, we came through Whigham, which has one main street with a single block of storefronts, charming and historic. For the occasion, the townspeople had hung colorful banners with an image of the trout lily on every telephone pole. The Whigham Diner and other businesses have added hours for visitors during the blooming period. The local community club has made T-shirts and comes out to the preserve to sell them and to volunteer. Whigham has adopted the lily with pride and now uses it as a source of identity for the town.

Whigham's annual Rattlesnake Roundup, a long-standing event, was advertised by a fifteen-foot banner suspended in front of the town's modest community club. Townspeople gather every year in March on the club's lawn to attend educational sessions about understanding and protecting rattlesnakes and to observe a few of the creatures up close and safely confined in an aquarium. Local vendors sell food, drinks, and crafts.

The diamondback is the largest rattlesnake in the world and can reach a length of six feet. The open-canopy pine savannahs of this region are its favored habitat, and the snakes have adapted to the use of fire in maintaining the forests by wintering in gopher tortoise holes and armadillo burrows.

Only a few years back, a very different practice at rattlesnake roundups took place across the region in early spring. Georgians brave enough to pursue the rattlesnake in the wild would hunt them down and kill them maliciously, sometimes by pouring gasoline into gopher tortoise holes. They would throw in a match to flush out or burn the creatures to a crisp.

Times have changed. The gopher tortoise is protected by state laws in Georgia and Florida, and eastern diamondback rattlesnakes are classified as "a species of conservation concern." The snakes have experienced a decline in the region because of habitat loss and killings. Fortunately, roundups like the one in Whigham have been transformed into educational events.

Back at Wolf Creek, visitors have arrived throughout the afternoon to see the lilies, whose daily blooms are beginning to fade. A couple from Jacksonville, Florida, drove three hours to get here, and they are ecstatic to see the preserve. A local Airbnb owner has brought her weekend guests to see the blooms.

Mysteriously, the peak this year is a week early, and the members of the Florida Native Plant Society and the volunteer leaders of the preserve have speculated about the reasons. Beth Grant says, "In 2018, we had three or four hard freezes, and it seemed like the lilies that year all bloomed at once. We'd never seen that before in thirteen years of observation."

This winter, there were two days when the temperatures dropped below freezing and stayed low overnight and into the next day, which some guessed might have brought the lilies out faster once the warming temperatures arrived. "We used to have one or two freezes that hit below twenty every year, but not anymore," Grant says.

"It is easy to say it's climate change, but we don't really know," Wilson Baker adds.

Since 2019, the Wolf Creek Trout Lily Advisory Board, made up of citizen volunteers, has been operating under the umbrella of Grady County government, and under an official vote by the county commission, the group continues managing the preserve. The advisory board was formally organized in October 2017 and is charged by the county with facilitating "scientific studies, removing invasive exotic plants, improving safe public access by means of improved or new walking trails, and guiding public and private groups of citizens through the preserve as part of an environmental education program." It's a good model for this unusual situation.

The volunteers have made small improvements every year since the preserve opened, and they have engaged the surrounding communities in the work to build awareness and support. Boy Scout Troop 383, from Whigham, has completed several Eagle Scout projects: a roadside marker and kiosk, and benches along the Wolf Creek trail. With help from donations collected in a jar at the head of the trail and grants from the state and the Resource Conservation and Development Council of Southwest Georgia, the managers have installed signs with QR codes along the preserve's system of pathways. Visitors can use their phones to learn about the medicinal plants and several other topics related to the local flora. Meanwhile, Dan Miller, the expert nurseryman, has managed to cultivate a quarter-million lilies from seed on his own seven acres in Tallahassee for further study. He has been a key professional voice throughout the negotiations with the county.

Christine Ambrose developed all the maps of trails and brochures. Early on, we learned, Dr. Ambrose directed her biology students at Thomas University in Thomasville to work on a digital map of the preserve with geospatial coordinates. Since her retirement, she has collaborated with staff at Tall Timbers and used some of their resources to help with the project.

Margaret Tyson said county officials have been surprised by how many people have visited and how far folks will drive to see the lilies. Some days, the dead-end road beside the tract has parked cars stretching a quarter mile down the lane on both sides.

The conservators are trying to strike a balance between keeping the preserve natural and accommodating guests. A portable restroom has been installed, and Tyson described plans to add a permanent shed so that volunteers can store some of the tools they need to remove invasive plants on the property, the signs they post during the spring, and the tables they use to put out brochures and sign-in sheets.

"We want to avoid making this a destination for other activities," Baker said. "If we were to put in picnic tables, someone might suggest bike trails, and the next thing you know, the property could take on another identity."

Scott Copeland, a preserve volunteer and member of the Florida Native Plant Society, fashions walking sticks made from sparkleberry, the largest native blueberry that grows here. Angus Gholson started this tradition by hand carving the sticks for friends he took on field trips to see the trout lilies. Though the blueberries are good eating only for birds, the wood of the sparkleberry is hard and can be burnished to a smooth and handsome shade of reddish brown. Most of the sticks Copeland makes now have a slight curve to them. His own favorite, made by Gholson, has just the right leverage to help remove invasives or dislodge a fallen limb when the volunteers are out clearing the paths each year, anticipating the month or so of visitors.

The preserve managers set out the fancifully named sparkleberry sticks for the use of guests as they hike the trails. They are also available to take home for a ten-dollar donation to the preserve. Donna and I generally keep our sparkleberry walking sticks in the car for outings back home.

Gholson called his informal group of hikers and wildflower enthusiasts the Sparkleberry Club. Betty Jinright, the woman who had scribbled the phone number of her philanthropic friend on a seed packet, was a member of the club. Betty is gone now, too. She died in 2019, at the age of eighty-seven. Her obituary said, "She was a floral designer beyond compare, and shared her talents with many."

Most of the managers of the Wolf Creek Trout Lily Preserve are retired now and will surely pass along their responsibilities to the next generation when the time comes. Let it be so, as long as the lilies bloom.

# 4

## River Otters

*"It'll be all right, my fine fellow," said the Otter. "I'm coming along with you, and I know every path blindfold; and if there's a head that needs to be punched, you can confidently rely upon me to punch it."*—Kenneth Grahame

Over the side of the canoe, the river is as clear as a fine goblet. A half-dozen dark gray fish, each about the size of my forearm, are heading upstream. "Mullet," says our guide, Sam Cole, who is following our clumsy canoe in his slender, neon-green kayak. Donna is in the front seat of the canoe with her camera. I am in the steering position, holding a paddle and a pen, with a notebook in my lap. Already a few drops of river water have blurred my swift scribbling.

The Ichetucknee River in the sandhills of north-central Florida is fed by nine named springs and multiple unnamed springs. It is five and a half miles long and forms the border between Columbia and Suwannee counties. The water temperature holds steady at seventy-two degrees Fahrenheit year-round, Sam says.

For the past twenty-eight years, Sam Cole has been the chief biologist in charge of monitoring the Ichetucknee's health. Like his father before him, he is a Florida park ranger. It was not difficult to persuade him to accompany us on our morning pilgrimage in search of river otter, a mammal I had never seen up close. "Any excuse to run the river is a good day at work," Sam said to me on the phone.

The closest sizeable town to the headwaters of the Ichetucknee is Lake City, the Columbia County seat, about fifty miles northwest of Gainesville, and home to some 12,000 residents. Historically, this section of Florida has been engaged in livestock ranching and row crop agriculture, with some mineral extraction conducted more recently. There was a definite cowboy vibe as we passed through Lake City early this morning. The presence of so many clear-water springs in the region has made Columbia County a destination for kayakers, cave divers, and snorkelers.

Sam kindly makes no comment as Donna and I—adjusting the bulky life jackets supplied by a local concessioner—pitch our canoe sideways to the current. Paddling Adventures, headquartered in the state park, provided the canoe and shuttled us to the put-in at the top of the run, where Sam was waiting. Because we launched so early, there were no other boaters, and we would have the river to ourselves for the entire trip.

The Ichetucknee is narrow at the headwaters, and stately live oaks, sweet gum, cedar, and cypress, all trailing beards of Spanish moss, arch over the water, creating a canopy. In places, the water appears to be boiling—the circular swirl driven by powerful underground springs.

The Head Springs, near our put-in, was for many decades the destination for community gatherings. As a swimming spot, it drew bathers year-round—even at Christmas. It was also the site of elaborate picnics and cookouts, baptisms, hymn singing, and many courtships and engagements. Local children learned to swim here, and fishing was always productive. Some partygoers brought their watermelons to chill in the springs; others discreetly brought homemade moonshine in jugs. They churned ice cream and would sometimes spend a whole day boiling peanuts in big pots over campfires on the banks.

In the 1920s, the community put up bath houses near the parking area and installed a long rope swing on the riverbank for those who dared to use it. More than once, according to oral histories collected in a book called *Old Timers Remember: Ichetucknee Springs*, automobiles parked too close to the river would roll into the water, in the early days to be retrieved by mules and, later, by wreckers.

As Sam puts it, "Getting in this river has been a local rite of passage for teenagers through many generations." Today, his job is to honor the local customs (which now involve large groups of young folks in inner tubes on the river) while protecting wildlife and water quality. The present rules are strict—no food of any kind is permitted on the river, and drinking water is allowed only in nondisposable carriers. Our driver to the headwaters made a careful inspection of our belongings. We noticed that there was no litter anywhere.

Donna and I bump awkwardly through the arch of a massive, partially submerged tree that was toppled into the water by high winds this spring. Sam makes a note to come back with a chain saw and a helper to make this new impediment easier to navigate. (To protect delicate underwater habitat, the first section of the river is off limits to inner tubes; only kayakers and canoes are permitted.)

Donna and I are still trying to get our paddles in sync. "I can't see what you're doing," Donna, who's facing forward, says. "Just be ready to shoot whatever you see; don't worry about paddling," I say. Sam is coasting on the current and setting up his eBird checklist for the day on his phone while we continue to adjust ourselves in the canoe. We, the landlubbers, have our phones in waterproof pouches around our necks—another vexation layered over the stiff Styrofoam life jackets. Sam has an instantly inflatable life jacket that lays flat on his chest like a black vest until needed. He's also wearing cool river socks and shoes. He has the legs of an avid bicyclist and the patience of Job.

I take a breath and try to take in the unfolding scenery. The sky is deep blue, but the water up ahead seems even bluer. No sign of river otters, however. Sam had hoped we might see some near the put-in. It's chilly yet, and little swirling dervishes of mist linger over the water, which is much warmer than the ambient air temperature. This slight fogginess enhances the mystery, and the river is meandering, so seeing very far downstream is not yet possible. The sun is low, and the shadows are long and angular.

A fish pops out of the river in a flash of silver and then hits the water with a sideways smack. "Mullet," Sam says. "We don't know why they do that, but some studies suggest the mullet may be trying to get rid of a parasite around their gills by hitting the water flat like that." Sam says that garfish are in the river, too. I wonder if we will see their impressive needle teeth in this clear water.

Below the schools of mullet swimming alongside us are underwater tresses of green eelgrass, waving languidly in the current. They are rooted to craggy deposits of ancient limestone, 40 million years old, along the sandy bottom. This part of Florida is dominated by karst topography, meaning it is a land of dissolving limestone bedrock that gives way to sinkholes, caves, disappearing streams, and underground springs. According to the United

States Geological Survey, these special physical characteristics account only for about a fifth of the country's landforms.

A USGS map of the lower forty-eight states shows that so-called sinkhole hotspots cover a significant area in Florida, from the Panhandle down to the center of the state. These are areas where the earth can open and sink down at random; the landscape is like Swiss cheese. Tennessee is also home to a concentration of sinkhole hotspots running up the Appalachian chain into Virginia.

Alabama, Georgia, and South Carolina have fewer sinkhole hotspots scattered about, but a great sweep of limestone bedrock runs up the coastal plain of these three states. The presence of calcareous rock—limestone and dolomite—is part of the reason for the wild South's enormous biodiversity. Certain plants delight in the mineral deposits.

Karst landscapes, with their underground aquifers and caves, provide critical drinking water to communities across the region. Once the caves gave ancient humans protection from storms. Karst caves and springs also served as the sites of early spiritual ceremonies of people who created cave paintings on the rock walls. Many of these archeologically important sites around the globe have been preserved and still draw visitors.

Some scientists have argued that karst topography could be useful for carbon sequestration and storage to combat climate change. Others disagree, suggesting that the destruction of natural carbon reservoirs through the extraction of calcareous rock (as in the process used to produce cement, for example) proves the environmental danger and damage of a strategy that would attempt to stow carbon in these natural rock "basements."

According to the German speleologist (cave scientist) Oliver Heil, "The biggest enemy of karst is the cement industry because they mine karst mountains, destroying caves and disrupting waterways." Heil, quoted in a 2021 article by the Heinrich Böll Foundation, a nonprofit environmental education group, goes on to posit that disrupting waterways, changing the path of water, and enlarging sinkholes will result in further erosion, landslides, and flooding, while putting botanical treasures and wildlife at risk. Currently, some 8 percent of global $CO_2$ emissions comes from the cement industry. The methods for drilling wells in karst landscapes or even under the ocean to sequester carbon are not energy efficient, and the danger of leaks has not yet been addressed, the article concludes.

Sam has told us that our ride down this river will be easy. There are no rapids. The elevation only drops about forty-two feet as the Ichetucknee meanders southwest and then merges into the tannin-dark Santa Fe River, which in turn flows into the Suwannee River. The Suwannee ends in the Gulf of Mexico near Cedar Key, a modest island community about sixty miles from here as the crow flies.

The spring-fed boils emerging from deep in the earth keep us moving along at a gentle pace. This swirling movement keeps the water clear— less susceptible to the leaching of tannins from leaf litter that falls into still water, sinks, and stays put, disintegrating and discoloring the water. According to a 2003 study by the Florida Department of Environmental Protection, "The depth of the Floridan Aquifer under Ichetucknee Springs causes more pressure where the river has eroded through the limerock, pumping an average of 233 million gallons of water daily." Unfortunately, more recent water levels in the Floridan aquifer have declined. Water is being drawn down for the use of metropolitan areas as far away as Jacksonville and Tallahassee.

Sam leads us around a bend and invites us to turn sharply into an area marked by a sign as "ecologically sensitive and closed for restoration." He explains: "We occasionally take early morning guided tours in these areas to interpret water quality and springs' habitat restoration, but normally these areas are closed to the public."

Our destination is the famous Blue Hole or Jug Springs, documented in an extensive story in *National Geographic* in 1999. It is 100 feet long and 25 feet wide, shaped like a jug, and the depth of the water in the sink hole below us is 45 feet. From where we sit, on the boiling surface, capturing the water's intense color and depth with a camera is impossible. Only photos from higher above do justice to this azure gem, set in a ring of trees silvered with Spanish moss.

The entrance to some 600 feet of underwater caves begins at the vent of this spring. Cave divers and snorkelers hike in to practice their skills here, and scientists come to learn more about the mysteries of karst topography.

Hiking out to Blue Hole on a hardpacked nature trail from the parking area is allowed, and a deck has been constructed for divers to stage their gear. This spot used to be the place where locals went skinny dipping, especially on New Year's Day. Now the hole is carefully protected, and the caves are accessible only by certified divers. Sixty-seven million gallons of water shoot from Blue Hole every day, making it what scientists refer to as "a first-magnitude flow." The thrust of the underground water is beguiling. It moves our canoe and Sam's kayak around as if our boats are dancing. Swimmers feel the push, Sam says, and find themselves getting a bit of extra exercise when they jump in and try to cross this cauldron.

It was here at Blue Hole that an aquifer mapping group proved that stormwater runoff from Lake City makes its way underground to Iche-tucknee. Dye released in the Rose Sink, a big sinkhole near Lake City, was traced to the Blue Hole. The Rose Sink operates like a bathtub drain, where trash and stormwater circle and are drawn down into the aquifer, traveling over eight miles to the Ichetucknee. Though the journey dilutes much of the damaging chemistry, there is growing concern about the presence of nitrates, phosphates, dissolved oxygen, and other pollutants in stormwater runoff—a common occurrence where agriculture is primary and where real estate development, with accompanying wells and septic systems, burgeons.

Ichetucknee Springs State Park's nearly 3,000 acres are not immune, though education and cleanup efforts have successfully engaged residents who treasure their long relationship with this river. Monitoring the presence of agricultural pesticides, coliform bacteria, fertilizers, gasoline, oils, and other pollutants from the fields around Lake City, along with urban stormwater runoff, is ongoing. Beyond the park's southern boundary, the last two miles of the Ichetucknee run through a private residential area where houses line the riverbanks. There, water quality is monitored by citizen scientists who volunteer with the Florida Springs Institute, a science-based environmental organization.

From Blue Hole, we turn back into the main waterway and soon reach Trestle Point, a section encompassing both sides of the river where phosphate mining was conducted beginning in the 1890s. African American workers for the Dutton Phosphate Company extracted hard rock phosphate by hand and shipped the ore to a nearby village on narrow-gauge rail cars. We see the large stone foundation for the trestle that once crossed the river here.

Another mining company, Loncala Phosphate, bought the mining property in the 1920s and allowed the public to enjoy the river, fencing off only certain areas. They did no extraction until the 1950s, when they reopened the spoil piles surrounding the old pits and scraped them to obtain phosphate residues. This effort, which ended around 1967, helped to clean up some of the damage and tailings from the original phosphate operation, Sam explains. Where Dutton Phosphate had dug clay-settling ponds, Loncala attempted to restore them by planting pine trees. Still, in the forest around us there are deep pits and areas where soil damage from the mining operation is probably best left undisturbed.

Here, I learned a new term. The disturbed lands in the park, in this case from mining and a long list of other human incursions, are called ruderal areas. Ruderal species are plants that are the first to recolonize disturbed grounds polluted by waste or refuse. The term comes from the Latin word for rubble, *rudus*.

The State of Florida finally bought this property in 1970 from Loncala Phosphate and established Ichetucknee Springs State Park. The area was designated a National Natural Landmark in 1972. The state performed some restoration to the Head Spring, where dumping had long been a problem. Back in the day, when land was cleared beside the river, the refuse would likely be bulldozed into the water. The state initially removed a lot of litter and waste from the Head Spring and river from 1970 to 1972.

"If I remember correctly," Sam said, "the final figure from the later cleanup project that ran from 1994 to 2008 was over 160 cubic yards of concrete, rubble, and scrap metals debris—enough to fill up a two-story, five-bedroom house."

Though still fragile, the current health of the river is remarkable, considering its history of human intrusion. A grist mill came along with the phosphate mining, and turpentine operations were also set up alongside the river, leaving scars on a few pines to this day. By the 1920s, the virgin longleaf pines had all been timbered. The present longleaf are all second growth. Slash pines were also planted in certain places as replacements. Cedars, Sam says, were cut for the manufacture of pencils in Perry, Florida. For their Christmas trees, locals cut small cedars that grew in the phosphate mines back then. Cypress was harvested for shingles.

In the 1930s, an entrepreneur launched a glass-bottom boat business, hoping to attract tourists. It was located near the site of a tavern, downstream from the present-day US Highway 27 bridge, but apparently some unhappy neighbors felled trees into the river to block the excursions, which mimicked a similar attraction at Silver Springs, Florida, that is still in business.

Dynamiting fish was a common practice in the old days. As one participant in the Ichetucknee oral history project confessed: "You could always tell when the fish had been dynamited. . . . The bones of the backs were always shattered." It took only a quarter stick of explosive dropped in a deep spot, and the perpetrators would scoop up their catch in a basket and have a fish fry on the banks. Moonshining in these parts was another undercover venture in the woods, and field hands regularly bathed in the river—clothes and all—after a day's work in nearby tobacco fields. In the 1930s, cattle entered the river even at the Head Springs and below the state park boundary. They were allowed in the river until 1980.

The number and scope of human disturbances to this river and the erosion from foot and vehicle traffic on its sculpted banks are hard to imagine in this floating moment in April 2023. But perhaps the most important factor in the contemporary restoration process has involved imposing limits on the numbers of people allowed on the river at any one time and managing access points and permitted activities.

Ichetucknee's renewal as a state park has allowed some species of concern to flourish, including the gopher tortoise, southeastern fox squirrel, southeastern kestrel, eastern indigo snake, and gopher frog. Controlled burns in the pine forests have enhanced the habitats of these animals and some rare species of plants. Even the Florida manatee will make an appearance here if the water level is high enough, Sam says. (They come up from the Santa Fe River, seeking food.)

River otters (*Lutra canadensis*)—the subject of a full shelf of children's books, many written in watery England—are the winsome and fleeting amphibious mammals we have come to see. They are regularly sighted here, though not in large numbers. River otters, like so many other fur-bearing animals, were harvested to the brink of extinction for their pelts in many regions of the United States in the nineteenth and early twentieth centuries. Their comeback is cause for rejoicing. Occasionally, as we move downstream, a fallen limb juts out of the water at just the right angle to suggest the slick furry head of a river otter, but no sightings yet. Still, I trust Sam's eagle eye.

The river begins to widen beyond Mission Springs, the site of a Spanish settlement established here in 1608. The mission settlement was preceded by the exploitive Hernando de Soto, who came through this area on his way west in 1539. His expedition journals detail the raiding of the Timucuan peoples' capital village of Aquacalyquen sited here. The Timucuan people later approached the Spanish in St. Augustine in 1596 for the placement of a mission; their delegation then returned and built all the mission structures before a Spanish priest took residence in 1608.

Later, in the waning years of the mission in the 1650s, Indian participation became forced labor. The ravages of European diseases that spread to the Indigenous people, and their rebellions against the presence of more and more invading settlers, eventually led to the abandonment of the mission by 1706. Archeological digs have located its footprint, however. Both prehistoric and historic artifacts have been uncovered in the park, including Indian mounds, but they have yet to be fully interpreted and preserved.

From the woodland hammocks we have passed through, the landscape now transitions to what residents refer to as the Grassy Flats, a wild rice marsh. Around a substantial bend in the river, the tree canopy opens, and we glide into bright sunshine. Native grasses stand tall along the shoreline, and prolific swamp lettuce edges the river. A stunning, black-crowned night heron flying overhead leads us forward, stopping twice in leafy limbs above us and then soaring again downstream in plain view. Among the taller grasses at water's edge, a snowy egret with its black bill, yellow eye, and those telltale feathers on the back of the head reminds me of a mullet haircut from the 1970s. The bird is fishing for breakfast.

A jet-black anhinga has perched on a tree snag to dry its wings in the sunshine, posing as the menacing villain in this bird montage—at least that's my anthropomorphic projection. White ibis are high up in a broad cypress like little mounds of snow, perched among the Spanish moss. Dozens of ducks swim around us, diving and teaching their young to feed —colorful wood ducks among them. Sam points out a pied-billed grebe, which is not a duck. Instead of webbed feet, the grebe has lobed toes and water-retaining feathers that allow it to sink and then dive surreptitiously for food. The river cooters and yellow-bellied sliders—all native turtles— are out in the sun, too, lined up on logs, heads held high.

An enormous osprey navigates its huge wingspan between thick limbs and then settles high in a live oak. As we float under it, its head and eyes are tracking us—a bit intimidating. We see a juvenile ibis, still sporting brown feathers on its backside and a bright white breast. Sam tells us to watch for a rare limpkin (*Aramus guarauna*) that looks somewhat similar.

The limpkin is a South American native, about the size of a juvenile ibis. Florida and southern Georgia are the northernmost edge of its range. So named for its limping gait, this is the bird that the irreverent Florida writer Marjorie Kinnan Rawlings shot, parboiled, stuffed, and roasted on a wood stove sometime in the 1930s or 40s. She served it proudly for dinner after an expedition on the Ocklawaha River near her home at Cross Creek, southwest of Gainesville. Rawlings, famous for her best-selling and heartbreaking novel *The Yearling*, about a Florida boy and his pet deer, did not realize the species was declining when she pulled out her .22 rifle, dropped the bird, and then waded up to her waist in the company of a water moccasin to retrieve her bounty. Rawlings declared the bird most delicious. To her credit, she eventually lost her appetite for wild game and adopted a more protective stance toward Florida's amazing wildlife.

Donna got a splendid photo of a limpkin on this section of the river, though we would not see just how good the image is until later. Nor did we hear its call, which has been described as haunting and banshee-like.

According to the Cornell Lab of Ornithology, "Males have especially long, looped tracheas, or windpipes, as cranes do. They use these to make loud, grating, piercing, high-pitched screeches and wails. During courtship, feeding, or when spotting a predator such as an alligator, Limpkins produce rattling calls." I wish we could have compared the limpkin's rattle to the sound of the sandhill cranes in Alabama. Sam informed us that a pair of limpkins nested there later in the season and raised four chicks.

Scientists have suggested that the recent spike in limpkin populations in Florida may be connected to the presence of exotic apple snails (*Pomacea*), also from South America, which limpkins love to eat. (One native apple snail and three exotics are now in the state.) These large snails eat from a broad menu of algae and will feast on such plants as cucumber, carrots, lettuce, spinach, and watermelons, along with dead fish and other snails. In turn, limpkins, raccoons, alligators, and otter will make a meal of the snails. Apple snails are sold in the aquarium trade as "housekeepers," since they will clean the sides and bottom of a fish tank, but they should never be released into a natural area like Ichetucknee. They damage native habitat and are considered a menace. "We have an abundance of the native apple snail species," Sam says, "but so far we haven't seen the nonnatives in the park—although they are in the regional waters surrounding us."

We paddle on. We come upon two alligators, both about four feet long, in this bird-dense area of the river. One is fully in the water, swimming with an above-water profile that could be drawn in single stroke—nose, eyes, and ridged back. The other alligator shows itself a bit farther downstream. I am shocked by how easily it climbs out of deep water to sun on a log. I had no idea they could navigate such a quick, vertical exit. I am glad not to be in an inner tube. We keep moving, steering a wide birth around the log. Sam says the few alligators in this river generally stay away during peak tubing season: too much splash and chatter from the humans.

Sam points out a platform where water quality monitoring equipment is stationed on the river. He paddles over and ducks under some decking to take a reading. His office is constantly collecting data and can get real-time measurements online from several stations along the river. But while we're here, it's good to put eyeballs on the gear, he says.

We finally reach the floodplain, the last section of the river within the park boundaries. Here, the banks seem lower, nearly at water level and punctuated with cypress knees and gigantic trees lining the water's edge. It is a cathedral of cypress. Sam is studying the shoreline intensely. This must

*A curious river otter pops up among the cypress knees in Ichetucknee State Park in north-central Florida in April.*

be the place. He sees a head pop up and then disappear. It is a river otter. He points. "There!"

I paddle us closer while Donna holds her long lens tight, ready to point in any direction. Sam pulls over downstream from us. The otters are cavorting between us in a tangle of swamp lettuce and vines strung among the cypress knees. They must have a den near here. The entrance to the den would most likely be underwater, Sam tells us, designed to keep out coyote, foxes, and other predators.

The first otter Donna captures with the camera shows only an eye and ear above the surface of the darkish water. Then come a few rapid-fire shots, showing only a swirl of water, and then, finally, she gets the full head punching up above the surface. We are staring down the nostrils of this otter, its nose level with the bulging eyes. Its bottom teeth are wholly visible and very white. This otter's protruding shape could be mistaken for a cypress knee but for its expressive face!

In the next photos she takes, the otter is giving the side-eye to something off camera—perhaps a mate or an offspring? Then the animal turns, and we have a side view of the head, full of glossy, dripping whiskers. It looks like Teddy Roosevelt without glasses! His little ears jut out just behind the eyes.

According to the National Wildlife Federation, a river otter's whiskers are highly sensitive and attuned to prey when the water is cloudy. The animals depend on their underwater vision when submerged, and because they are nearsighted, they rely more on their sense of smell, touch, and hearing to navigate above water. Watching them, it is easy to believe these animals are performing, playing with us, aware of our curiosity, and unafraid.

Then the swimming commences. Sam thinks the parents of this family are heading across the river now toward the other shore, leading us away from their young somewhere still under the quilt of swamp lettuce and vines.

"Did you hear them barking?" he asks. We nod. The sound is squeaky. Sam told us earlier about witnessing a mother otter climb out of the water with a fat sunfish in her mouth. Her pups watched, whined, and squeaked as she ate the entire fish alone, sharing not one bite. "They have to learn how to feed themselves," Sam said.

Donna and I turn the canoe upstream and paddle against the current to the far side, following the otter parents. We pull close to the shore and into an eddy where we can sit still for a bit and watch.

It takes a while—otters can remain underwater for up to eight minutes without a breath—but suddenly the two adult otters are taking turns popping their heads up through the floating greenery on this side of the river. It's like whack-a-mole at the state fair, and we are laughing. One comes up and climbs fully out of the water, onto a log, slithering and slick, oblivious of the cooter that has established dominion on the far end of the log. The otter seems to be watching us. Then, it has another idea and curls its long body in a perfect upside-down U and disappears again under the foliage and the water.

Otters are three to four feet long, and a third of that length is the strong tail. They weigh between eleven and thirty pounds, and the males are generally larger. They have webbed feet and sharp claws.

Kin to weasels, minks, skunks, and badgers, the river otter has two layers of fur. The first is the waterproof outer coat, which appears slick, almost oily. Under that is a layer of finer fur. Then comes a layer of fat that keeps

*A North American river turtle, also known as a cooter, shares a log and some sunshine with a voracious river otter who has been eating plump crayfish caught in the Ichetucknee River.*

the animal warm. When the otter goes underwater, the fur becomes silvery with air bubbles clinging to it.

Once, these acrobatic creatures densely populated North American rivers, but an aggressive fur trade in combination with loss of habitat and water pollution reduced their numbers drastically by the beginning of the twentieth century. According to Jeff Traynor of the Furbearer Conservation Project, over the past fifty years, the reintroduction of river otter has been a successful initiative in many states and on tribal lands. By 2016, more than 4,000 river otters had been translocated to twenty-three states. More carefully monitored and managed harvesting continues for the fur trade.

Now, here comes the other otter, slightly smaller, climbing onto the log with its mouth full of crayfish. The orangish tail of a crustacean is protruding from the otter's jaws, at least for a second, before the otter chomps down and chews noisily, crunching the shellfish. Donna is getting it all, frame by frame. This show is like the old joke at the dinner table: "What are you eating?" Mouth opens: "See food!"

The crawfish tail is now hanging from the otter's mouth like a cigar, but its moments are numbered. The otter swallows and shows us her empty mouth. She turns to sniff at the back end of the static turtle, then turns back to dive under the foliage. We can see two undulating forms burrowing under the swamp lettuce like puppies playing under a bedspread. We are so close to them that Donna wonders out loud if they are going to swim under the canoe and dump us over. We laugh. We can't help it. But the feasting goes on, the otters come up with mouths full, and more crayfish violently perish. Finally, the turtle turns to observe the otter at the end of its log. So much commotion, it seems to be telegraphing with despair.

One of the old-timers in the Ichetucknee oral history book is quoted in the early 1950s as saying: "The crayfish were so thick you couldn't see the white bottom of the river." We get the impression that this site on the Ichetucknee is just as plentiful today.

The otters, sated, finally head upstream, leaving a swirling wake behind them. We are feeling as full and happy as they must be. Sam is grinning from across the river. "Did you get what you wanted?" he shouts. We nod enthusiastically.

We are quiet for a time before reaching Coffee Spring, marked by a sign and cordoned off from river riders because it is the sole home of the Ichetucknee siltsnail (*Floridobia mica*), also called the sand-grain snail, so tiny that it resembles a grain of sand. This aquatic gastropod lives only in this ten-square-yard habitat in the spring pool. It is seriously threatened but holding its own for now. The snail feeds on algae, bacteria, and other microbes. We pause to consider what it means to be visiting in the only place on the planet where a small family of a singular creature survives.

"Some things are better, and some things are worse," Sam says of his twenty-eight years here. "Algae bloom has increased a lot. It used to be that a column of water from the river was perfectly clear, but now we are seeing 250,000 people on the river over the course of a year." He pauses. "The wildlife has improved. The beaver are back." (*Ichetucknee* means "land of the beaver.") "We also see more turtles and alligators and otters. Just the other week, I saw nine otters in a day." Right now, I am counting ten turtles in my line of sight, ancient black creatures that seem like custodians of the rocks around us, watching and waiting.

As we prepare to leave the river, Sam points out one last highlight, a new beaver dam on the far shore. He will leave us now to paddle back upstream to the north end of the river, returning to park headquarters. He says he recorded more than forty species of birds on his eBird checklist for the day.

In 1999, a corporation aimed to put a cement plant within four miles of the Ichetucknee River, and citizens rose up and opposed it vehemently. The controversy inspired the newly elected governor, Jeb Bush, to take a float trip on the Ichetucknee, which in turn led to a budget-line item for springs protection in Florida. Although that funding ended in 2010, the science-based organization the Ichetucknee Alliance is still working to protect the river. Florida is home to more first- and second-magnitude springs than any other state in the nation. They are a marvel worth protecting.

# 5

# Alligators

*My life would never again be the same. Indeed, whoever has beheld the manifold charms of this paradise of woods and waters, must come away fascinated and spellbound. Its majestic pines and cypresses, its peaceful waterways, and lily-strewn prairies, together with the wild creatures that inhabit them, should have been safeguarded forever from the lumberman's despoiling operations.*
—Francis Harper

"All the alligators you see here are real," says the sign at the entrance to the Okefenokee Swamp Park, a venue that sits just outside the northern rim of the 438,000-acre National Wildlife Refuge, near Waycross, Georgia. What does it say about us as twenty-first-century human beings that such a message is necessary? This is the largest blackwater swamp in North America, where it's still possible to go back in time and encounter authentic wilderness. Thousands of American alligators (*Alligator mississippiensis*) live in the Okefenokee National Wildlife Refuge and the adjacent Swamp Park. It is hard *not* to see one.

As we would soon learn, the warning signs date back to the era of Oscar, once the dominant alligator in this vicinity. Oscar was well known to regular Swamp Park visitors over several decades. According to a story in the *Florida Times-Union*, Oscar "was considered docile around visitors, some of whom were surprised when they approached what they believed was a fake alligator until he was roused from one of his many naps on the grass or walkways near the gift shop." Reportedly, a woman once sat down on Oscar's back, believing him to be a sculpture, and she lived to tell about it. Now the organization that runs the Swamp Park is taking no chances.

Oscar's territory, experts say, probably ranged over as many as 100 acres, where he entertained his "girlfriends" and sired many alligator hatchlings. According to a retired park educator and naturalist, Don Berryhill, Oscar was billed during his time as the largest known alligator in Georgia. He was thirteen feet long and weighed a half ton, according to the lore. He died of natural causes in 2007 and was close to eighty years old.

Dr. Kimberly Andrews, a coastal ecology specialist with the University of Georgia's Marine Extension and Georgia Sea Grant program, says it would be impossible to provide an accurate count of the thousands of alligators in the Okefenokee because the refuge is vast and impenetrable in places. "And baby alligators are purposefully well hidden by their mothers," she says. "That is the essence of a refuge. When humans can get to every single place where animals are, that's when we start having a conservation problem, right?" Kimberly is the supervising scientist who manages the alligator research that is supported by the nonprofit Okefenokee Swamp Park, Inc.

To get here, we had driven south from the town of Waycross through Dixon State Park on Georgia Highway 177, a rough, two-lane road that stops just short of the refuge proper and picks back up only at the western entrance to the swamp. It was April, and we were charmed by the profusion of hooded pitcher plants (*Sarracenia minor*) rising out of the sedge on both sides of the road. Nearly 98 percent of pitcher plant habitats have been destroyed in the southeastern United States, so seeing these striking, federally endangered specimens was a rare treat. Tall with reddish hoods, they are an indicator of just how poor the soil is here and how rich the insect environment. (Pitcher plants have evolved to trap, dissolve, and take nourishment from insects rather than nutrients in the ground.) Squint and these outlandish plants could be mistaken for long-necked birds.

Coming in, we also examined the dainty lavender blossoms of pickerelweed. Their long green stalks are an indicator of the acidity of the water they live in. The plants, which have fan-shaped leaves and dense clusters of stems, share a symbiotic relationship with the pickerel fish, providing them cover from predators in the Okefenokee's black water. Other species that thrive here include the toothy bowfin, flier (a pan fish), warmouth, and the bullhead catfish.

While we admired the flora, a swallowtail kite (*Elanoides forficatus*) flew over, gliding on the breeze and hunting for prey. These deft raptors, with their scissor-like tails, appear to be dressed in black-and-white formal wear. Native peoples believed they were messengers from the world of the sky to earthbound creatures. Their wingspan is more than four feet, and they will eat large insects while flying. In breeding season, they may also capture frogs and lizards to feed their young. Though once common throughout the Southeast, kites are now rare enough outside Florida that birdwatchers in other southern states will organize excursions to find them. The Okefenokee is a migratory stop for kites during their annual winter journey south. What an introduction to this isolated Eden of resilient survivors in harsh territory!

Finally, we arrived at the historic amusement park at the end of the paved road. The Georgia Forestry Commission holds the lease for the Okefenokee Swamp Park and Okefenokee Adventures. Both concessions are managed by a nonprofit organization that is more than seventy-five years old and governed by a board of trustees. These venues offer ongoing educational programs to visitors, including guided tours, boat rentals, and field trips. They also support academic research designed to increase understanding of the complex biological ecosystems and cultural significance of the Okefenokee.

Recently, Kim Bednarek, the executive director of the nonprofit, has been working closely with the US Fish and Wildlife Service and the Department of the Interior in Washington, DC, to fulfill a new regional vision with the National Wildlife Refuge at the center. It's an effort to protect the swamp while also engaging in a form of rural economic development that's sustainable.

The partner organizations have obtained official notice that the refuge is on course to become a UNESCO World Heritage Site. It will be the first US Wildlife Service property to do so; many hope the designation will open the way for additional funding for this and other wildlife refuges. Already, the swamp is recognized as a Wetland of International Importance by the Ramsar Convention, and it was named a National Natural Landmark. The eighty-acre Stephen C. Foster State Park, on the west side of the refuge, has been identified as an International Dark Sky Park, one of fewer than 200 essential stargazing sites so named on the planet.

"Being declared a World Heritage Site," says Bednarek, "will put Southeast Georgia on the map internationally and on track to allow residents of the area to lead more prosperous lives while protecting our unparalleled natural resources. We are working to help our communities get ready for the international attention and the positive placemaking that will come from this designation while continuing to protect the Okefenokee. A refuge, by definition, puts wildlife first. That won't change."

The Swamp Park facilities at the north end of the swamp have long served as a gateway for visitors with young children who might not be ready for some of the more demanding natural aspects of the environment. Here, young families can take a noisy and smelly diesel train ride through the woods or a short boat excursion on a circular canal. There are natural history exhibits and a small collection of wildlife in captivity —mostly turtles and snakes.

In the gift shop, we saw Oscar's image on T-shirts and ball caps. Oscar's skeleton, carefully assembled in a glass case, was a sobering sight—our first up-close introduction to alligator anatomy. Don Berryhill and his colleagues dissected Oscar after his death. Besides the grasses, rocks, and bones that they discovered in Oscar's stomach, the team found evidence, still visible in the skeleton, that Oscar had been shot with both a pistol and a shotgun during his long life—not an uncommon finding in the remains of alligators that have survived such attacks.

Outside, on the far bank of the canal, Donna found a live alligator to photograph. Later, I sent the image to Mark Hoog, a South Carolina native and recent master's-level graduate of Georgia Southern University. Over several years, Mark and his colleagues have captured more than 300 alligators in the Swamp Park and have tagged 174 with transmitters so they can follow the animals' habits and whereabouts using satellite telemetry. Mark knows these gators up close.

"Most likely, that is Obadiah," he told me after examining the photo. "Given the size and the location, that is a favorite spot of his to bask." Obadiah is one of Oscar's successors in the role of dominant male in the park. In his work, Mark has followed Obadiah and many other Okefenokee alligators, examining their behaviors across different family territories. He's also documented their genetic relatedness by taking DNA samples from individual animals.

Mark and Kimberly work in tandem to capture alligators by using a treble hook to snag the hide. They reel the reptile in using a heavy-duty rod. It is no small feat to snare the animal with a looped cable and then tape its mouth shut with electrical tape before bringing it onto a boat or onto a bank. They use a towel to cover its eyes to calm the creature and wrap the tape around its jaws while sitting on the animal's back. They attach rope restraints to the hind legs while taking measurements and blood and tissue samples to assess the alligator's health.

From these data, Mark has been creating an alligator family tree for the Okefenokee. The researchers have proved that Oscar's progeny moved out of this immediate area after his death. Only one of his heirs has been positively identified here. Following Oscar's reign, there was Crazy, who lived for a time in captivity in the Swamp Park. Then came Okefenokee Joe. Both alligators were old and died after a short time. Now Obadiah is the king of the territory, and Sally is the dominant female. Sally has been followed through many breeding seasons so that the researchers can learn more about maternal behaviors and the alligator lifecycle. Another graduate student, Kristen Zemaitis, from the University of Georgia's Odom School of Ecology, has studied alligator diets, finding that they eat most anything that moves, including some unexpected invertebrates, such as moths.

The research component of the nonprofit mission will continue even as plans call for a future in which visitors to the Okefenokee can experience a deeper and more comprehensive interpretation of the swamp's natural history and cultural significance, whether they enter through the refuge's north, west, or east gateways.

The name "Okefenokee" originates from a Native American word that means "land of the trembling earth." This great sunken bowl of unstable ground is deceptive. What looks solid may not be. Islands of sphagnum peat have swallowed up people and wildlife over the centuries and created a menacing reputation for the swamp.

The last time the Okefenokee broadly engaged the popular imagination of the American public was through the comic strip *Pogo*, created by Walt Kelly. The strip ran from 1948 to 1975 and featured a philosophical possum and alligator, along with a broad cast of other talking animals from the swamp.

Earlier, in 1940, the Georgia novelist Vereen Bell wrote *Swamp Water*, a fictional account of the hardscrabble life of the white swampers who drove out the Native peoples and settled here before the 1930s. Bell's story was first serialized in *The Saturday Evening Post* and then became a best-selling novel, still in print. It was made into a film in 1941 by the French director Jean Renoir and was reprised in another Hollywood production in 1952 called *Lure of the Wilderness*. A talented storyteller, Vereen Bell was killed in action at the age of thirty-three in World War II. Of the Okefenokee, he wrote: "The swamp was a lost world; a nether land where the crawling things in the muck and the screaming things in the air had triumphed, and man rotted in the peaty earth; the space was the space of eternity, endless, changeless; a thousand years passed, and another thousand, and another, and man's bones were ground to ash by the ages, his gun a discoloration of rust beneath the earth."

Don Berryhill, who became the keeper of Oscar's legend, was a pharmaceutical representative who moved to Waycross in 1965. He began working on contract with the swamp as a science educator. "When I was kid up in North Carolina in 1941, I went to a movie in Concord called 'Swamp Water' about the Okefenokee Swamp," he told the *Florida Times-Union* shortly before his retirement from the Swamp Park at the age of eighty-two. "There's a scene that stuck in my mind I can still see, a man sinking in quicksand."

Incredible tales of the people and wildlife in the swamp are a Georgia tradition. I made my first visit to the swamp in the 1960s from my hometown of Atlanta. It must have been just before Mother's Day when my father cooked up the idea of a weekend road trip for my grandmother (his mother), my brother, and me. I might have been in the seventh grade. We would drive the 250 miles to Southeast Georgia, where Dad had reserved a cabin in the Stephen C. Foster State Park for an overnight stay. He'd scheduled a guided boat trip the next morning.

We four carefully balanced ourselves in the flat-bottomed johnboat with its small Evinrude motor. Our guide had also brought a long pole to push us in shallow water. I remember weaving our way out into the swamp. I hoped to see alligators and *not* to see water snakes dangling from overhead trees. My tiny grandmother, Stella—in her seventies then, and wearing her usual cotton dress and Keds—was expert at confronting water moccasins, though she had left her special hoe reserved for this purpose back home. (I was raised to fear snakes—cottonmouths, especially—but now I am learning how they are essential parts of this ecosystem, as are alligators. Both are graceful swimmers.)

Not long ago, I asked my brother what he remembered most vividly from that trip more than fifty years ago. His response was the same as mine would have been. It was the sound of the swamp. We arrived after the long drive near dusk, and once outside the car, we were immediately deafened by a raucous chorus of frog voices—a 360-degree symphony from the depths of the understory and the towering cypress and pines around us. We had to shout to hear each other over the ruckus.

Now Donna and I aimed to repeat that experience with special attention to the alligators—probably the closest creatures we have to the ancient past in the wild South. We booked accommodations in a new complex of rental cabins on the west side of the park, near Fargo, Georgia. From our screened porch, we heard a few frogs calling as evening fell. There were some wet spots in the surrounding lawn—frogs need moisture—but the day had not been warm. It was surprisingly quiet that night.

We had reserved seats for a pontoon boat tour, leaving early the next morning from the docks in Stephen C. Foster State Park, which is located very near the headwaters of the Suwannee River. (The Pittsburgh song-writer Stephen Foster never saw the Suwannee River, by the way. He picked the name out of an atlas and dropped the *u* to create his famous lyrics.)

I'd read an essay by the Georgia naturalist Janisse Ray about paddling white-knuckled for three days all the way across the swamp, through the maze of wilderness water trails that are mapped out for those who dare. Knifing quietly through the black water without a motor was appealing, but the effort required is not for the faint of heart. The swamp covers more than 600 square miles. The combination of landforms and waterways—lakes, narrow streams, plant-covered channels, floating peat batteries, and prairies—is daunting.

Our motorized trip began through a narrow canal that runs into Billy's Lake, in the main channel of the Suwannee. Our guide seemed tentative, clearly new to the job, but his six passengers were enthralled. We were quiet at first, dazzled by the birds, insects, and alligators in their teeming habitat.

The first few gators we saw that morning were basking, immobile, in partial camouflage among water lilies (*Nymphea odorata*). The yellow lily buds were emerging but not yet open. The lily pads encircled the dark bark of countless swamp tupelos with their bell bottoms sticking out of the water. We also watched yellow-rumped warblers and a dapper kingfisher up high on a limb, and we heard a white-eyed vireo calling, a sound both musical and raspy. Rows of Virginia chain fern (*Woodwardia virginica*) waved at us from the shallows. Thick resurrection ferns (*Pleopeltis polypodioides*) clung to the stocky arms of passing water oaks.

"We have three types of egrets and seven kinds of woodpeckers in the swamp," our guide announced. "Red-bellied, red-headed, downies and hairies, yellow-bellied sapsuckers, pileated, and red-cockaded." The ivory-billed woodpecker, now believed to be extinct, once lived here, too, I had read. A specimen of the bird from this swamp, taken in 1913, is in the permanent collection at the Academy of Natural Sciences of Philadelphia.

We watched transfixed as the pontoon delivered us around a bend. There, a little blue heron (*Egretta caerulea*) with its purplish head feathers was standing bowlegged on a raft of waxy green lily pads. The bird's long bill was clamped on a sizable frog. With binoculars, I could see that the frog's head was already obscured, almost in the bird's gullet. The frog's legs were limp and hung straight down, as if they had no kick left. Its forelegs stuck straight out like the arms on a crucifix. The heron's wide eye, seen from the side, conveyed the challenge at hand—how to swallow a meal longer than its neck.

*A stoic alligator rests among lily pads near Billy's Lake in Georgia's Okefenokee Swamp.*

Long ago, the white swampers, who lived in a world of their own, would have called this heron a "blue scoggin," according to the ornithologist Francis Harper (1886–1972). He spent many years documenting the local biota and was taken with the inhabitants' quirky names for their world. They called ospreys "fish eagles," the screech-owl a "scrich owl," and the Carolina wren a "fence dodger." Though he was not formally trained as a folklorist, Harper faithfully recorded the swampers' tall tales and the lyrics of the songs they sang at community frolics.

Harper talked to many a swamper who described days when the alligators were so thick in the pools and lakes that a human "could have walked across on 'gator heads." They also told of coming upon alligators thrashing up the lake bottoms to force fish to rise higher in the water. On these occasions, the larger water birds would wait at the shoreline to feed themselves on the smaller bounty that the alligators did not consume themselves.

People feared the alligators but also hunted them for their hides. One storyteller said he collected fifty-six skins in a single outing. The reptiles are federally protected now, and Kimberly Andrews says they still stir up the water, drawing fish to the edges of the swamp where birds will seize a quick meal.

A passenger on our boat asked what time of year is best to visit the swamp. The guide thought for a moment and then spoke. "Well, spring break is when the weather is good, and the bugs are tolerable," he said. "We have extreme weather in the summer. We won't go out in the boats at all if there's lightning." For the record, Okefenokee Adventures recommends fall and spring as excellent times for guests, but summer is prime time, too. Summer visitors might want to consider wearing long sleeves against the insects, however.

The guide then explained how a wildfire in 2007 took out 500 acres of the swamp near Waycross to the northeast. "It was natural, from lightning, and the peat burned up. It was bad. Cypress grows slowly—one inch takes thirty years. And forty percent of Georgia's GDP comes from pines," he added. He pointed to a grove of bay trees in the distance. "They are a member of the gardenia family with a magnolia scent," he said, but we didn't get close enough to pick up the perfume. The number and variety of trees around us were magnificent.

Another passenger asked about the surprising fullness of the red maples so early in March. "They started budding out in January," the guide replied. "Last month, the blue flag swamp iris were blooming, too. They smell like Fruit Loops." We laughed. He then pointed out damselflies strafing an alligator on a log on one side of the boat. He cut the engine, and we drifted for a time.

There was no place in sight friendly enough to consider disembarking. The tangles of titi (*Cyrilla racemiflora*) and greenbrier (*Smilax laurifolia*) were thick and wicked looking. Never mind the alligators. They seemed so primitive and heavily armored—something like oversized hand grenades lodged in the muck. But when they were floating free, half obscured in the water, the gators seemed benign, their big eyes clear and thoughtful, their jaws closed in a permanent, upturned smile.

We heard voices behind us. A group of canoers soon overtook us. Their boats were loaded with gear—tents and provisions. They explained that they were on their way to spend the night at a shelter—one of thirteen that have been constructed above the land and water, deep in the middle of the swamp. "Your paddle will be used every inch of the way as you wind through cypress forests or cross open prairies exposed to the sun and wind," a refuge brochure explained, with only a hint of the strenuous nature of the undertaking. My friend, the naturalist Liz Domingue, wrote a guide to the Okefenokee, and she had already reminded me that to know the swamp, you have to see more than you can see on a pontoon boat ride. "There are so many habitats," Liz said. "Wet prairies, shrub islands, hammocks, small open lakes, black water channels. It takes at least an overnight in the swamp to appreciate the vastness and variety of the place."

The first humans in the region mostly traveled into the swamp to hunt game and to fish, entering from the St. Mary's River to the southeast or the Suwannee on the west side. European settlers later were intimidated by the swamp's thick wilderness and wrote off the place as unproductive territory or a big blank space on the map to be circumnavigated. The presence of malaria-carrying mosquitoes, biting yellow flies, diamondback rattlesnakes, water moccasins, bears, panthers, cougars, wolves, and alligators made this silty and sunken place seem forbidding.

The first Europeans to venture swamp life settled along the edges in cabins, but around 1850 a hardy group of families moved farther in and stayed for generations, enduring the hardship of isolation while reveling in the rewards of such bountiful hunting and fishing. They developed a profound sensitivity to their natural environment, according to Francis Harper's affectionate narrative in the book *Okefinokee Album*, which was collected and edited by Delma E. Presley after Harper's death.

Like the Dismal Swamp, on the border between North Carolina and Virginia, the Okefenokee was used, for a time, as a refuge by Native people avoiding the European invaders. Later it became a pathway to Florida for some enslaved Africans escaping their captors in South Carolina and Georgia.

Outsiders considered the idea of commercially developing the swamp just prior to the Civil War, but preliminary surveys came to nothing. Nevertheless, the conversation had begun about the possibility of draining the waters, collecting the peaty soil, and putting the cleared swamp to agricultural use. According to contemporary estimates, some 200 million metric tons of peat —serving as a huge carbon sink—have been accumulating over the past 6,500 years in the Okefenokee. It is the largest contiguous, intact freshwater peat deposit on the North American coastal plain, according to Rena Ann Peck of the Georgia River Network. Thank goodness it has remained so.

In 1890, when the Georgia legislature opened the swamp to bids, an entrepreneurial group calling themselves the Suwannee Canal Company purchased the land for 26.5 cents per acre and set about to build a drainage canal. Their fundamental misunderstanding of the hydrology of the terrain stymied their success. They did manage to harvest valuable cypress before the banks of the canal collapsed under heavy rains. Then a battery of hurricanes hit the region just before the turn of the twentieth century. The company folded.

A second group bought the swamp out of receivership in 1901 and surveyed the terrain more thoroughly, developing a piecemeal plan to harvest more cypress and pine over time. Eventually, after various sections of rail lines being used to carry out the lumber sank in the muck, the company landed on an idea to install vertical pilings to hold up the train tracks. They soon found that some of the thirty-foot pine pilings driven into parts of the swamp never seemed to hit solid ground. The poles disappeared in the muck. Despite these setbacks, the company arranged to hire and pay local swampers to extract a significant amount of timber before the Great Depression halted operations altogether.

Francis Harper's brother, Roland, a quirky botanist and civil engineer, championed the renewable resources and scientific value of the Okefenokee. A contemporary historian and writer, Megan Kate Nelson, analyzed the moment: "Scientists began to enter the Okefenokee just as three movements coalesced in American culture: industrial logging, conservation, and professionalization of the sciences."

The Harper brothers and colleagues of theirs at Cornell University worked hard to promote the treasures of the Okefenokee. Members of the Georgia Society of Naturalists escorted several US senators and others on field trips to camp and fish in the swamp. Francis wrote vividly and passionately about the swamp in the magazine *Natural History*. It didn't hurt that his wife, Jean Sherwood Harper, made a personal appeal to Franklin Delano Roosevelt on behalf of preserving the wilderness. She had once served as a nanny to the president's children, and FDR listened. Ultimately the Okefenokee was designated a National Wildlife Refuge in 1937. Today, the swamp brings some 300,000 visitors each year to the area, contributing significantly to the economic well-being of this section of Georgia.

Before he turned the boat around and retraced our route back to the visitor's center, our guide veered into the Middle Fork of the Suwannee on Minnie's Run. Here the understory thickens, the river narrows, and the canopy closes in. Cypress trees tower above holly and wax myrtle thickets. Flowering plants called golden club glowed like lit candelabras hovering above the water. Each plant sent up tapers that ran from pink to white, held firm by thick green leaves. Each of these "candles" was topped with a golden spike, as if a flame were burning at the tip. The black water heightened the elegance of the display.

Francis Harper, on his first encounter with this plant in 1912 on a boat trip guided by the veteran swamper Dave Lee, called the golden club by a different name: "The luxuriant blades of 'never-wet' (*Orontium aquaticum*) in the water almost shut out a view of the surface, and they rustled and scraped along the sides of the boat." "Never-wet" refers to the sturdy, water-repellant leaves of the plant.

Ultimately, we heard no frogs or bellowing alligators on our pontoon ride; we only witnessed that silent victim—the frog that the blue heron among the lily pads ate for breakfast. Back at the boat launch, we decided to visit the Suwannee River Sill, a destination on Highway 177, leading into and out of Foster State Park. This area can be accessed without any admission fee. The dirt road to the impoundment, marked clearly with a sign, comes just before the Fish and Wildlife ticket booth as you are heading into the park from the west side.

When drought and wildfires damaged a half-million acres in the swamp and beyond in the mid-1950s, the US Army Corps of Engineers began building this earthen dam, known simply as the sill. It was installed to keep water levels higher in the swamp in the event of drought, with the aim of fire prevention.

Alas, the dam made no significant difference in water levels across the vast swamp. It has now been breached, and the floodgates are left open. It's still a popular fishing spot, though the consistent presence of alligators makes the angling a little tricky. The alligators are adept at stealing fish that have been caught and are being brought to shore by folks using a rod and reel.

"Those alligators know exactly how far they need to be out of the range of the hook," Kimberly Andrews told me later. "The gators don't risk getting snagged, but they position themselves perfectly to grab a caught fish off the line. The people that regularly fish the sill are used to it. Some say the alligators ought to get their share. This is their place, too."

Knowing that we are in the alligators' home place is critical knowledge here. It is a fundamental lesson that the swamp teaches, and you must pay attention. A fishing lure landing in the shallows or a splashing fish being reeled in will attract the attention of alligators. "It seems sensible to find a better place to fish than where alligators are numerous," Kimberly suggests. Kids who are interested in dinosaurs will likely be awed by alligators, and the Okefenokee is a great place to see them. Just don't let children or pets get close when you visit.

A narrow, paved lane tops the earthen dam. A few wider pullouts are positioned along the length of the road so that oncoming cars can pass each other. The dam provides a means to view the alligators that seem to prefer this quiet spot. You don't even have to get out of your vehicle.

On our first visit, the giant reptiles had stationed themselves on both sides of the channel that runs parallel to the dam. Here, freshwater runoff from the river backs up above the dam. Juvenile brown and mature white ibis were feeding in the shallows of the channel, high-stepping along, unperturbed by the bumpy-backed beasts the color of gray muck that were sunning only a few feet away. Adult white ibis struck dramatic poses among the gum trees on the far side of the water. Down the channel, a few alligators had taken to the water. Though Donna stepped out of the car to shoot images of this ancient tableaux of kindred creatures, I was more cautious. It is less than twenty yards downhill from the pavement to the channel.

The alligators seemed serene and unperturbed. Perhaps that's why some people can't resist taunting them with food. Some even throw marshmallows, Kimberly said. Not smart. Humans feeding alligators for fun is a kind of animal persecution, as she put it. "When alligators grow accustomed to associating humans with food, they cannot distinguish between a pet and a chicken leg someone threw out to them," she said. Alligators tend to hang around the edges of a body of water, and it is illegal to throw anything at them or to harass them.

Alligators are ectotherms, meaning they must regulate their body temperature through their environment. They take to the water to cool off in warmer months. When it's chilly, they bask in the sun to stay warm and will stop eating when the temperature drops below seventy degrees. When temperatures drop below fifty-five degrees, they will go dormant and build a burrow for wintering. They can go for long periods without food. "A 100-pound dog will eat more in a year than an 800-pound alligator," Kimberly said.

Though alligators are known to vocalize—with bellows, deep growls, and burps—they also make low-frequency sounds that humans cannot hear. Mark Hoog said that gators can live for sixty to eighty years in the wild. Male alligators average 500 pounds in weight and can reach twelve feet in length. "After that, they tend to add more bulk than length to maintain their dominance over other males," Mark said. "An alligator's girth can reach five feet."

Females will produce twenty to fifty eggs in a clutch. About 10 percent of these eggs become hatchlings that reach maturity, Mark explained. In the Okefenokee, bears, raccoons, coyotes, and some other mammals will raid alligator nests to eat the eggs. From wild-game cameras positioned at alligator nests in the Okefenokee, the researchers have seen female alligators defending their nests against bears and chasing off turtles.

"Turtles will lay their eggs in an alligator nest," Kimberly explained. "Alligator nests are huge—five feet across and two or three feet tall. If the turtle can get in and out and lay her eggs without conflict with the alligator, then it's good to go," she said.

They investigated Sally's nest one year, and when they were excavating, they found baby turtles hatching out. Apparently, the fierce mama alligators consider the baby turtles their own if they make it into the nest. "It was really cool," Kimberly said. "We call that free childcare for the turtles. Especially if the alligator is a dominant female. They are often the most vigilant in defending their nests."

Immature alligators eat insects, amphibians, smaller fish, and other invertebrates. Adult alligators eat snakes, turtles, small mammals, fish, and birds. As top-level predators, alligators are prime candidates for absorbing mercury and other heavy metals from contaminated food. Mercury in the Okefenokee habitat could be passed up through the food chain and concentrate in larger animals.

Not so far from the Okefenokee, at the Savannah River Ecology Lab in South Carolina, where Kimberly worked for a time, scientists have conducted decades of research on the impact of changing water temperatures and the presence of certain pollutants on alligators living below the Savannah River nuclear site, which is on the border of Georgia and South Carolina. In the Okefenokee, says Kimberly, the presence of mercury is confirmed but does not yet seem to be concentrated enough to affect the health of the mature alligators—at least not yet.

The alligators on the banks of the sill never moved discernibly that day. I thought I caught an eye blink or two, but I wasn't sure. To witness a movement takes fierce concentration and focus. They were so big and inert. I alternated between staring and snapping pictures of them with my phone.

The tails on the largest creatures took up half their length and looked impossibly strong. Their legs are oddly short compared to the body mass. However, knowing that they could suddenly move up to a speed of thirty miles an hour if stimulated, I stayed attentive. Seeing so many alligators at once without any barriers was a powerful meditation on human vulnerability and the scope of evolution that's still taking place on the planet.

You don't have to master the complex family trees of alligators, crocodiles, and dinosaurs to feel in your human bones the connection of these reptiles to the prehistoric creatures that once traversed the earth. As the Savannah River Ecology Lab explains it, "Alligators and their relatives are the last of the living reptiles that were closely related to dinosaurs, and their closest modern kin are birds."

The bird connection became more vivid to the researchers when they witnessed Sally facilitating the feeding of her babies. In the video they made of this scene, Sally's four hatchlings were stationed in the shallows against the bank in the Swamp Park. Nearby, Sally had dug out a spot for herself in deeper water, facing away from her young. Without moving her head at all and without propelling herself forward, she stirred the mud with her back legs, and then gently swept her long tail gracefully back and forth, harvesting invertebrates and pushing tadpoles and minnows toward her hatchlings as they grabbed at breakfast, looking as eager as baby birds in a nest.

We returned to the sill thirteen months after that first visit, and the gators were lined up only on the far side of the channel. None had apparently come across to the dam side that day, but they were aligned differently on the far bank. Instead of being parallel to the channel, they were perpendicular to the edge of the water, and a few had their heads partially submerged. As we drove the length of the dam, we counted them—sixteen in all. A couple of the gators had their mouths wide open, showing their seventy-plus teeth jutting out of the fleshy red tissue inside their jaws. Again, they scarcely moved, and I wondered if they were expecting lunch to land in their mouths. I learned later that this behavior is a means of cooling off in the same way that a dog pants. They orient themselves to the sunlight.

Francis Harper conveyed a tale told by a swamper named Thrift, who observed an old alligator in the swamp that held his mouth open until his tongue was covered in mosquitos. "He'd slap his mouth together like that. . . . He'll chew a little bit, it look like. An' then he'd open up again an' keep repeating that. Every time his tongue would get covered, he'd make another bite."

Such was entertainment in simpler times at Okefenokee.

We were unsettled to learn that alligators are cannibals. They eat each other if food is scarce or if they are defending territory. Sometimes older siblings will even go after members of their mother's next brood. We saw a hint of this aggression during our second session on the sill, when an alligator with its mouth frozen open moved the rest of its body in a deliberate way to suggest that the next alligator should move no closer.

Kimberly told me that male alligators will often fight as if they were in a dive bar, and sometimes gators die in these fights, but they don't always become dinner afterwards. They are killed simply for invading another male's territory.

As we reached the spillway at the end of the dam, an enormous osprey suddenly flew up into a tree. Apparently, it had been fishing in the rushing water being released below the dam. The osprey was massive, and my heart was thrumming at the surprise. Just then, a woman came driving into the turnaround area in a van. She rolled down her window and told us she was looking for a place to camp for the night and somewhere to walk her dog. The small dog, old and feeble, was sitting in her lap behind the wheel. We strongly recommended that she find another site to spend the night. Remarkably, she then asked if we'd seen any alligators. I guess she hadn't looked to her right while crossing the sill!

Alligator skin became popular in the 1800s as a form of leather. Like the river otter, the Carolina parakeet, and the white ibis, all harvested for fur or feathers, alligators were killed in great numbers in the United States to feed the fashion industry. By the middle of the twentieth century, some local protections for alligators were introduced, but poaching continued. With the creation of the Endangered Species Act of 1973 and several other pieces of federal legislation, alligators finally received protection that had some teeth, pun intended.

Like the river otters in the last chapter, alligators have made a historic comeback, though they still supply food and leather to the marketplace and are farmed for these commercial purposes. They are generally bred and harvested in captivity from eggs collected in the wild. These days, hunting and alligator farming are tightly regulated and taxed to under-write conservation efforts for the species. Alligators are still gravely threat-ened by habitat destruction, pollution, and their similarity in appearance to crocodiles, which are often killed if perceived as a threat to humans.

Earlier in her career, Kimberly Andrews worked intensively to help coastal communities in South Carolina and Georgia understand the value of keeping alligators as predators within an ecosystem and how to live alongside reptiles, including snakes, with respect and care.

Alligators are important players in their ecosystems. They enhance their wetland habitat by digging "gator holes" in the marsh, providing refuge in deeper and cooler water for themselves and for other creatures when the landscape goes dry. For this contribution, alligators are designated a key-stone species, meaning that without them, other species suffer.

When alligators or their habitats are removed from a local ecosystem entirely—a practice that is accelerating in many coastal developments across the South—the consequences are serious. "Raccoon populations explode," says Kimberly. "People say, *Where did all these raccoons come from?* And they have no idea. The same happens when snakes are removed —explosive rodent problems develop."

Andrews and her colleagues at the University of Georgia's Marine Extension and Georgia Sea Grant program encourage developers, civic officials, and neighborhood groups to adopt an informed and compassionate policy of cautious cohabitation with alligators and snakes because of their contributions to wildlife management and ecological balance in coastal developments.

Still, the threats continue, even in the Okefenokee Refuge, established almost a century ago.

As of this writing, an Alabama firm called Twin Pines proposes to mine titanium and zirconium on an ancient sand dune that is less than three miles from the southern edge of the Okefenokee Swamp. Titanium is an ingredient used in toothpastes, sunscreen, and paint. Titanium alloys are also used in aircraft and weapons systems.

According to the Georgia Conservancy, "The proposed mining project would mine for minerals (titanium dioxide and zirconium) at a depth of fifty feet below the ground surface, which is below the level of the Okefenokee Swamp depression and integral to maintaining surface water and groundwater hydrology in this region of southeast Georgia. Twin Pines plans for a facility on a larger 12,000-acre tract along Trail Ridge and very close to the Okefenokee National Wildlife Refuge in Charlton County."

Scientists have expressed concern that such a mine adjacent to the swamp could permanently damage the hydrology and fragile ecosystems in the refuge and disrupt the peat deposits that naturally sequester carbon underground. Others have said that mining can be a short-term disruption of habitat, which can sometimes be successfully restored. For this reason, they say, a focus on mining is misplaced and depletes energy and resources that conservationists should spend to resist the encroachment of commercial developments adjacent to the refuge, which can destroy critical habitat, introduce invasive species, and degrade water quality.

"Locating a mine too close to the swamp is a threat, and the wholesale clearing of land for permanent housing developments and manufacturing plants worries me, too." Kimberly Andrews says. "Any project that creates significant impermeable paved surfaces will increase stormwater runoff, which can affect water quality and quantity in the Okefenokee." The water quality of the swamp has already been degraded by the presence of mercury and other heavy metals and pollutants that enter as rainfall.

According to the latest statistics from the University of Georgia Marine Extension and Georgia Sea Grant, the diversity of ecosystems in the Okefenokee is profound. The swamp encompasses an assortment of more than 620 plant species (including four carnivorous plant species), 39 fish, 37 amphibian, 64 reptile, 234 bird, and 50 mammal species. The alligators may bellow and fight, but they can't stop this kind of potential damage to this one-of-a-kind place on the planet. People must take up the challenge and recognize the costs of habitat loss.

With such persistent threats to the Okefenokee, I asked Kimberly and Mark about the rewards of their work with alligators and what might be ahead for them.

"Honestly, the biggest moments of success are actually some of the smallest," Mark said. "When you're getting through to someone, and you're making them think differently about alligators, when you can see the 'ah ha moment' for them, that's rewarding. It comes when people begin seeing alligators and snakes as creatures that have important roles in the ecosystem, rather than being just some big scary predator."

"And even when you have a success, it's not permanent right?" Kimberly added. "You have to keep fighting for it, because something new will happen. Someone's going to come in and want to change things. Maybe you stop alligator removal in one place, but the pressure from human development and activity is unrelenting—even in our protected areas like the Okefenokee. That's why conservation is a career and not an initiative. We don't get to stop."

Not enough citizens have been convinced that some places need to belong to the alligators, the lily pads, the little blue herons, and the yellow flies. We can't possibly run it all, but we can ruin it all. As Walt Kelly's Pogo said, "We have met the enemy, and he is us."

# 6

## Frogs and Toads

RICEBORO, GEORGIA, AND CARRBORO, NORTH CAROLINA

*Frogs have an inelegant way of taking off from invisible positions on the bank just ahead of your feet, in dire panic, emitting a froggy "Yike!" and splashing into the water. Incredibly, this amused me, and, incredibly, it amuses me still.* —Annie Dillard

A few years ago, Erin Cork, a Georgia biologist, was relocating to the coast and looking to buy a house. A real estate agent took her out in the countryside to look at listings south of Savannah. As they drove toward one particular house in a rural area, the agent confessed that the location might be too remote and the air quality not so good. It was near a water treatment plant. The women arrived at the address, got out of the car, and approached the back door of the house. It was then that Erin spotted a spadefoot toad.

"It was just a tiny one," she said, "maybe a couple weeks old and clinging to the back step." Erin immediately began investigating the yard, not the house. "The realtor was just standing there on the porch in her capris and heels, staring," Erin said, grinning. "She had no idea what to do with me."

Erin bought the house. It sits in a bog. When the temperature is above fifty degrees Fahrenheit, the frogs there are so plentiful that their sticky toes cling to Erin's windows in the night, pursuing insects that are drawn to the lights inside. When it rains, Erin has an instant, outdoor laboratory, perfect for frog observation. So far, she's counted fifteen frog species, an impressive number for a single yard, she says. And yes, there are snakes (a dozen species seen to date), salamanders, and toads, and moths by night in the surrounding forest. "The undeveloped nature of the landscape, the extensive swamps, and managed forests are what drew me and the frogs to this area," she said.

I met Erin at the Coastal WildScapes Symposium, an annual environmental education event in Richmond Hill, Georgia, where I was giving a book talk. The conference theme was "Connecting People through Nature." In the afternoon session, Erin offered a crash course in frog identification with photos and recordings she's made of various species in her neck of the woods.

In her early forties, with wild red hair and a wry sense of humor, Erin is an entertaining teacher. Her descriptions of frog sounds are memorable and useful to anyone who wants to undertake the sensory adventure of identifying frogs by their music.

Frogs sing during the breeding season. The timing varies among species, from February to August in many parts of the South. If there has been rain in recent days, Erin explained, the chances of hearing frogs increase. But no matter what, you needn't travel far to experience frog music in the wild South. Chances are, you can find a thrilling (or trilling) chorus of frogs in a nearby park or bog if your timing is right.

"Some frogs will call intermittently during the day, so you can pick up clues in the daytime for your later explorations at night," Erin said. "But a half hour or more after sunset is when you want to be out there. You won't hear some frogs outside of a two- or three-month period each year, and some frog species may call only on a few nights a year, holding out for near-perfect weather conditions. You should try lots of different nights."

Erin founded the Coastal Georgia FrogWatch Chapter—a volunteer organization that monitors frog populations in her region. In her day job, she's a wildlife biologist with the Georgia Department of Natural Resources Wildlife Conservation Section. She's been involved with amphibian conservation projects since 2013 and is active in the Georgia Plant Conservation Alliance. She also serves on the board of the Coastal WildScapes organization.

Some years back, while earning her master's degree in wildlife ecology from the University of Georgia, Erin studied the habitats and management of rare amphibians in longleaf pine communities, focusing on gopher frogs, the rarest frog in Georgia. (As we learned in the chapter on trout lilies, longleaf pine forests are a fragile and diminishing habitat, and that's where gopher frogs live.)

According to herpetologist Whit Gibbons, one of the authors of the definitive guide *Frogs and Toads of the Southeast*, there are more than forty native and introduced species of frogs and toads in the Southeastern United States. Because some 20,000 species of amphibians are in decline worldwide, frogs in the United States were once monitored with federal support, but with budget cuts to the North American Amphibian Monitoring Program, volunteer citizen scientists have stepped up to help fill the gap in data collection.

Launched in 1998 by the US Geological Survey and now managed by the Akron Zoo in Ohio, FrogWatch is sponsored by the Association of Zoos and Aquariums in partnership with the National Geographic Society. Through FrogWatch, small groups of amateur naturalists go out in the field and record the types of frogs they hear in the wild. These nationwide data are collected online, mapped, and analyzed using a platform developed by National Geographic to determine vulnerabilities in frog populations.

In Georgia, five different FrogWatch chapters, including the one Erin started, are at work across the state. Alabama and South Carolina have five chapters each. Tennessee has four, North Carolina seven, Florida eight, and Virginia eleven. If you want to get involved in your state, Google Frog-Watch USA to find the group nearest you. Each chapter has a coordinator, but each member is critical to the observation process. The teams make excursions into natural sites at different times during breeding seasons to capture data.

Erin joined FrogWatch when she was still living in the university town of Athens, Georgia. In that group, she says, "We had a great mix of frog-gers who showed up regularly—including a new birder and a retired plant ecologist. One had an eye for spotting arboreal treefrogs, and the other was good at noticing interesting insects and plants. Another FrogWatcher, a filmmaker, was adept at spotting salamanders poking their heads out of holes. I was the only amphibian biologist in the group, but we all contributed equally to the experience."

Erin described the proper protocol for frog listening to her audience at the Coastal WildScapes gathering. It's simple, she said. A small group —usually no more than seven people—will head to a designated wetland. They might monitor only one wetland or several along their regular monitoring walk. They stop as a group and remain quiet for two minutes, giving the frogs a chance to forget the initial disturbance of approaching footsteps. Then the listeners cup their hands around their ears for three minutes, motionless, and listen carefully to distinguish and identify the breeding calls they hear. They will already have had some coaching in identifying sounds through their FrogWatch orientation training. If the listening is interrupted by a foreign noise—the slap of a beaver tail on the water, the roar of an eighteen-wheeler driving by, or the arrival of another creature, Erin said, they repeat the procedure from the beginning, always making notes about what they hear. They also bring headlamps and flashlights to spot frogs, but not before they've done the critical work of listening for them.

Frogs are a sentinel species. When you hear frogs, you are probably hearing a healthy wetland habitat. If instead there is silence, then something may be wrong in the environment. As Erin explained, "Some species begin calling when the temperature is in the low forties, but anytime it is fifty-five Fahrenheit and above—with decent humidity or recent or approaching rainfall—I expect to hear somebody calling."

Much to my surprise, I found a frog orchestra and chorus in my own backyard in the North Carolina piedmont a couple of months after I got home from the Coastal WildScapes conference. The frog sounds were coming from the bioretention pond in my Carrboro condo complex, where I live when I am not in the Blue Ridge Mountains. It was an April evening. Dark was coming on, and the pond had become the stage for a full-blown frog opera. I recorded the divas with my phone and sent the audio clip to Erin for species identification.

The dedicated purpose of the bowl-shaped, ephemeral pool in this small housing development is to receive the runoff from our semipermeable parking lot. The pond was designed to mitigate erosion and contain all rainwater on our common property. Our progressive, green-minded town mandates this water retention.

The pond also prevents stormwater from washing down a steep slope at the very back of the property, which would swell and muddy up the little urban creek at the bottom of the hill. We regularly see deer and foxes coming out of the woods here. We also hear hawks by day and owls by night, even though we are only about two-tenths of a mile from the busy center of town with its de rigueur restaurants, organic and vegan grocery, and purveyors of beads, backpacks, craft beer, and CBD oil.

The retention pond is loaded with granite riprap where the water comes in from the storm drain. This rock slows the flow. The bowl is also mulched and planted with natives—purple iris, pussy willow, and river birch: species that don't mind having wet feet. It's apparently the ideal temporary pond for our local frogs, and I'm sure the owls and hawks that sometimes perch on the swing set near the pond find easy meals below in the greenery. Examining the proximity of my own habitat helped me to realize that this sensory adventure with frogs is a simple, do-it-yourself experience, perfect for this book. You may not have to travel far, even in an urban environment.

With Erin's analysis of my recording, I learned that Cope's gray treefrogs and southern toads were singing their hearts out that May night, along with a few alarm calls from green tree frogs. I asked Erin about the alarm calls.

"I think, in that situation, the green tree frogs were just hanging out there on the vegetation," she said. "It's probably a great place for them to go forage and find bugs. Vegetation in the pond likely provides several different microclimates for frogs to bask or take cover as needed while foraging. Green tree frogs only breed in more permanent pools," she added, explaining that the alarm call is something they do if they sense some other creature coming near the edge of the pond.

So, when the green tree frogs are not issuing alarms, what do they sound like? I asked. Erin laughed. "With a whole chorus of green tree frogs, the individuals could have slightly different pitches. You might think you're hearing different species, but you're probably not. I heard just one or two calling at your pond." Erin said that breeding green tree frogs issue a syncopated sound. She imitated for me the beats of the rhythm—kind of a *cha-cha, cha-cha cha*. Suddenly I wondered out loud if Erin had studied music as a kid.

"Yes, I did," she said, a little surprised by my question. "I played piano for a really long time and studied music theory." She paused to think about it. "So yeah, that does help in my work, I guess." Erin went on to explain how researchers in her field use spectrograms—the visual signature of sounds—to analyze frog voices. "We review spectrograms on computer monitors so you can easily compare the audio recording to the image, if needed. There are even call recognition models that scientists have developed to run audio files through to help identify the species without the researcher having to listen to hundreds of hours of frog choruses."

I found the rhythm of the calls in my backyard concert compelling. Knowing of my interest in frog songs, a friend who lives high up in the Appalachian Mountains of North Carolina, near the borders of Tennessee and Virginia, sent me a video of the water feature that her husband built between their house and the driveway. Their plan was to have Japanese koi swimming in the stone pond, which they could observe from the kitchen windows above. The local raccoons and herons had another idea and cleared the pond of the ornamental carp in record time. In place of the fish, and for several years now, wood frogs (*Rana [Lythobates] sylvatica*) take over the pond and erupt like an agitated raft of ducks every February, even as the weather is still quite frigid at 3,500 feet on the mountainside. Erin explained to me that small ponds without fish provide the best habitat for many frog species that can't coexist with fish predators, so my friends provided an unintentionally perfect habitat.

The first time I heard the wood frogs quack when I was visiting there in person, I thought a flock of young ducklings had just landed. The sound is raucous and loud, and the eggs these wood frogs leave in the water are huge, viscous masses. Though the top layer of eggs sometimes freezes over, there are enough eggs to keep this procedure going every year, apparently.

The frog sounds in my own yard by comparison were more of a drone, akin to the whine of insects, reminding me of summer nights at my grandmother's house in Georgia before anyone had air conditioning. With the casement windows rolled open, I'd go to sleep with the whir of cicadas and the deep *glup-glup* of bullfrogs down at the lake, dreaming of the vegetable and dahlia gardens my grandfather planted on the bank. I could visualize the tadpoles stirring in the shallows, trying to avoid becoming dinner for the largemouth bass that my cousins loved to reel in from those murky waters.

The most common frog sound you are likely to hear throughout the South comes early. The spring peeper (*Pseudacris crucifer*) is a chorus frog, a type of tree frog, which is easily camouflaged on tree bark. Many times, in our travels for this book, Donna and I would slow down on a country road, lower the car windows alongside a low place in the terrain, and listen to a peeper chorus, which can be heard most any time of day or night.

Peepers were studied extensively throughout the twentieth century, but only recently has interest spiked in understanding how the timing of their singing and breeding may offer clues about climate and temperature disruptions in particular locations. "Our spring peepers down South get going in the winter, with some eager frogs calling as early as the fall," Erin said.

The late Gary M. Lovett, of the Cary Institute of Ecosystem Studies, wrote about the vocal power of these tiny creatures: "Males situated on vegetation overhanging water produce a loud 'peep' call, repeated typically fifteen to twenty-five times per minute," and they can produce "a call as loud as songbirds that weigh ten to a hundred times as much." The sound of their choir is intense and giddy. They are also distinguished by a dark X on their backs: hence the Latin name *crucifer*, or "cross."

Peeper frogs are only an inch long, so finding them is best achieved by listening for their music. They feed on ants, beetles, flies, and spiders. Eggs hatch into tadpoles in as little as two days. Tadpoles become frogs in six to twelve weeks. They have short lives—three to four years at most—according to the National Wildlife Federation. Most people have heard spring peepers, and the calls of other frogs in our region are extremely variable and fascinating, once you do a little research with your ears.

Here are some of the southeastern frogs you might hear, described by Erin's creative, species-identifying similes, along with a few accounts of their sounds and behaviors, as explained by other frog experts:

The pinewoods treefrog (*Hyla femoralis*) breeds in temporary wetlands. A roadside ditch will do just fine, thank you. It's found nearly throughout Florida and in lower Alabama, with some habitat reaching as far north as Birmingham. It can also be found up the coastal plain from Georgia through Virginia. A pinewoods treefrog seems to be sending out its signal in Morse code, says Erin, if you are old enough to know what that sounds like. Listening to a recording of this frog, I can see in my mind's eye the metallic tap of the code key and hear the static beep of the signals transmitted. Morse code assigns a series of taps—long and short—for each letter of the alphabet. Telegraph operators spelled out words, letter by letter, with these taps. They sent messages by wire strung across the country before the turn of the twentieth century. Later, Morse code was sent through wireless radio broadcasts from the trenches during World War I and from military ships and airplanes during World War II.

In contrast to the pinewoods treefrog, the barking treefrog (*Hyla gratiosa*) sounds like a pack of hound dogs off in the distance. They are the largest native treefrogs in the Southeast, with a length reaching more than two inches. (Unfortunately, the nonnative Cuban treefrog is now the largest treefrog in the Southeast. This invasive species can gobble up many of our smaller native treefrogs, Erin said.)

The barking treefrog can be found in the coastal plain and all the way up through Alabama and into middle Tennessee. The online guide from the Savannah River Ecology Lab at the University of Georgia—hereafter referred to as the SREL guide—tells us "males generally call from the water." By appearance, they resemble "tennis balls as they float, inflated, on the surface of the wetlands." Pictures of these creatures are amazing.

The squirrel treefrog (*Hyla squirella*) sounds like a raspy duck, Erin said. Unlike the wood frogs found in the Appalachians, these frogs are found up and down the coastal plain of the southeast and are only an inch to an inch-and-a-half long. Much like chameleons, their bodies can assume different colors, going from solid green at one moment to tan with dark brown mottling in the next, according to their environment. They will often come toward porch lights in search of insects, says the SREL guide.

The southern leopard frog (*Rana [Lithobates] spenocephala*) makes a sound like "a finger rubbing against a wet balloon," as Erin describes it. With spots like a leopard, these frogs are active day and night and are widespread in the seven states covered in this book. They can be found both close and far away from water. As the SREL guide declares, "Often a heavy winter rain will prompt explosive breeding in this species." Erin says they also do a great alarm call that's like a squeal.

"You'd think if you were trying to get away from a potential predator, you wouldn't make so much noise and then jump into the water with a big splash. But it's effective!" Erin said. "Another defense against predators that seems counterintuitive to our minds is the bright yellow or orange 'flashing' that many treefrogs have concealed along their thighs. When the frogs detect a threat, they leap away, exposing a bright flash of color, which is thought to startle or confuse a would-be predator."

Most sources I consulted agree that the southern cricket frog (*Acris gryllus*) sounds like glass marbles hitting each other. This frog is small and bumpy, with a distinctive triangular marking on its head and a stripe down the back. It comes in all assortments of colors. It has no toe pads and does not climb well, so it tends to stay along the edges of ponds, lakes, or slow streams in the coastal plain. The northern cricket frog (*Acris crepitans*) is like its southern counterpart but is found in the piedmont and mountain sections of the Southeast. Notably, the northern cricket frogs tend to click their marbles faster. And as the name implies, both southern and northern species might be misidentified as crickets.

The little grass frog (*Pseudacris ocularis*) makes a tinkling sound, almost like fine crystal. It is the smallest frog on the continent and has a black stripe running through its eyes and down the side of its reddish-brown body. It's an eastern coastal plain resident and is found almost everywhere in Florida.

Cope's gray treefrog (*Hyla chrysoscelis*) is the one I heard in the bioretention pond. It trills a single note that can be roughly imitated by the human tongue rolling an *R* sound, almost like a cat's purr but sharper and higher pitched. "It's a shorter, buzzier trill than that of the more musical southern toad," Erin said. It's one of the most common frogs heard in early summer all over the region covered in this book, except for southern Florida. Copes like to hang out high in trees. These two-inch frogs are best left untouched and appreciated only for their sound, however. They produce a toxic secretion through the skin that can be painful to humans.

Then we come to the toads. I asked Erin to describe the difference between a frog and a toad. She said, "I don't like that question, but I'm glad you asked, because there's really not a good difference between frogs and toads." Erin explained that the words "frog" and "toad" originated in England, where there were only four known native species at the time. "So, they called two of them frogs and two of them toads," she said, rolling her eyes. By the British definition, Erin went on, a true frog in the family Ranidae had long legs and usually moist skin and was very closely associated with water throughout its life. "True toads," in the family Bufonbidae, usually come with warts or bumps and with poison glands on the tops of their heads. These British toads, as well as toads around the world in the true toad family, have short, stubby legs, and they move very differently from frogs—in a series of short hops, rather than expansive leaps. Toads were considered more terrestrial and less aquatic, except during breeding periods. But, said Erin, as scientists around the world have studied amphibians more closely, they have found a much more diverse taxonomy. "The true toads and the true frogs are only two of six families of frogs that you can find in the Southeast, and two of thirty-six families of frogs found in the world. We have this crazy abundance of frog species, and you can see so many unique traits and behaviors that can't just be cleanly split between the classifications of frog or toad."

The gopher frog is a prime example of how the two original classifications don't work. "The gopher frog lives underground like a toad would," Erin said, "and it's got dry skin, but it certainly leaps like a frog and does a lot of other things that are what we think of as true frog. So, in Britain they didn't have the biodiversity to get the language right the first time. It's much more complex, and different amphibian species have adapted to very specific conditions and are still adapting."

The eastern spadefoot toad (*Scaphiopus holbrookii*) lives underground and can stay there for a year or longer during periods of drought. Its comical, low-pitched call sounds impudent and whiney, somewhere between a cat and a sheep. Instead of a meow, it's more like *maow*, and it trails off. "What stands out about our spadefoot toads," Erin explained, "is their large protruding eyes, with vertical pupils like cat eyes and the 'spade' feature on their back legs that allows them to dig backwards to excavate tunnels several feet underground." This is the species that sold Erin on her cottage near the water treatment plant.

The southern toad (*Anaxyrus terrestris*) sounds to me something like those metallic noisemakers you twirl at a New Year's party, only higher pitched. "Most describe the southern toad's call as sonorous—a long, sustained musical trill," said Erin.

*This eastern spadefoot toad, photographed by biologist Erin Cork in her backyard in Georgia, has bright-yellow catlike eyes with elliptical pupils—a likely adaptation to a life lived mostly underground.*

The eastern narrow-mouthed toad (*Gastrophryne carolinensis*) sounds like "a bus full of sheep," Erin told her audience at Coastal WildScapes. The recording she played of this toad had us all laughing at the precision of that description. Erin told me later that this species is also prevalent in her yard. "They're strange little frogs," she said. "They are triangular or teardrop-shaped, and they're pretty flat. They have a narrow, pointy little mouth, and they also have a fold in their skin—kind of like they are wearing a turtleneck—and they are ant eaters! You can imagine why that fold of skin might offer a little bit more protection while they are laser-focused on an anthill with ants streaming at them."

Depending on your age, some of the similes used to describe these frog and toad sounds may or may not be familiar. When going out to scout frogs, why not ask your children or grandchildren to make up their own descriptions or similes for the sounds? The point is to distinguish among the different species, eventually know their names, and remember their fine and varied voices. They are all different instruments in a fabulous orchestra.

The Patuxent Wildlife Research Center, of the US Geological Survey in Maryland (mentioned in the first chapter as the refuge where Canus, the whooping crane, lived), has developed a useful frog identification website. It allows anyone to look up frog species by state, common name, and scientific name, and to hear recordings of each species listed. This website could be helpful in preparing for your first frog excursion. Or you can easily record frog sounds on a smart phone and then take them home and use this guide to match the sounds you heard. Search online for "USGS Frog Quiz."

I realized that once I began to identify frogs, I got better at paying attention to everything that makes noise in the woods. Add in the cicadas, katydids, crickets of summer and fall, the occasional owl, and the songbirds of the season, and you've got an interesting auditory challenge and a means to increase your sensory capacity. Frog music could even lead a child to an interesting career like Erin Cork's. She got her inspiration, she says, from her father.

Erin's dad, who had a career in the military, once caught a kingsnake in the yard when the family was stationed in California. Erin was maybe four years old. She remembers how calm and gentle the large glossy snake appeared to be in her father's arms. It wasn't long before she became the snake handler in the family.

"When my father was stationed in Fort Leonard Wood, Missouri, they allowed my brother and me to run a veritable zoo in our unfinished basement. At any given time, we would have an assortment of birds, frogs, snakes, turtles, and rodents that we kept in various terrariums, dog kennels, and buckets. These were animals that we would find and observe for a while before releasing them back into the wild. My mom was never a snake fan, but she allowed it because she assumed (incorrectly) that snakes could not climb stairs!" Erin said. "As an even younger child, I remember filling a bucket with snails and leaving it uncovered in the garage. The following morning, my mom was horrified to find the walls and ceiling of the garage covered in snails." Erin smiled with mischief.

"I've always been interested in creatures that I could get my hands on or observe up close," Erin continued. "I really didn't bother with birds because they were so elusive, and you couldn't really watch them for very long. I think frogs are so accessible."

As the Corks moved to other postings, including Germany, Erin spent hours in the woods. She was never afraid of snakes and continued to be fascinated by them. Georgia would be the state where the family finally landed.

"Back then, we didn't have the internet," Erin said, "so I couldn't look things up with the same kind of fanaticism that I could later on." The idea of a career in natural history and conservation didn't occur to her for many years, however. Erin majored in English in college. After earning her degree, she worked as a bank teller, as an editor at a magazine, at a health spa, at a music venue, and at a nonprofit organization. "I was all over the place," she says. When she took a position as a college counselor at the University of Georgia, she found herself helping adult students who were returning to school. "It was definitely inspiring to see all these people ready to change their lives and find new careers," she said.

As an employee of the university, Erin was eligible to take classes for free. "I started taking wildlife classes on my lunch breaks," she said, "trotting across to the other side of campus." Eventually one of her professors asked why she didn't just get a degree in the field. "I'd worked a couple of wildlife jobs, and I really loved it. So, I started the graduate program. Eventually, I began my career as a wildlife biologist—a much delayed start, but better late than never."

I asked how she keeps herself going when the news of species extinctions, climate crises, and so much loss of habitat is around her every day.

"Sometimes I just don't pay attention," Erin confessed. "I have to go off on my own and have my own experiences—cool, outdoor wildlife encounters. That's a selfish thing for me. But my work week to week can sometimes be demoralizing. It's not as bad for me as what some biologists encounter. I am fortunate to work with private landowners, and a lot of them are really awesome—doing impressive, generous work that they don't have to be doing."

Erin helps landowners who seek to conserve wildlife and wildlife habitat on their land. They often partner with the state to protect rare, threatened, or endangered species. She said, "In Georgia, more than 90 percent of the land is privately owned. Compare that to land ownership patterns out West, where most of the land is owned by the federal government and includes large national parks and wildlife refuges, and you can quickly see why wildlife conservation in the Southeast needs buy-in from private landowners." Erin conducts baseline surveys and reports on properties with rare populations; she then connects landowners with financial incentives to promote and implement conservation management activities on their land—procedures such as prescribed burning or wetland restoration. She provides technical assistance and recognition to landowners with high priority species or habitats, and she helps secure permanent protections through land acquisitions, conservation easements, or cooperative agreements.

Among the large landowners she works with are families, and some are timber and power companies that voluntarily cooperate with the state to support wildlife conservation on company lands through the Forestry for Wildlife Partnership of Georgia's Department of Natural Resources, a program Erin oversees.

Erin also serves as an agent in negotiations with landowners who want to capture solar energy. Solar farms convert agricultural fields and forest land into big installations of solar panels, which are becoming more and more prevalent in the South, given the long days of sunshine in the region. Solar panels block the sun from the ground and heat up the landscape, interrupting wildlife corridors and accessibility. While solar energy is good for the environment in many ways, there are still trade-offs when it comes to land conservation for plants and wildlife.

"Longleaf pine and sandhills in particular have been targeted by solar developers because they are high and dry and can be less expensive than agricultural land," Erin explained. "We've got new solar developments that are in direct conflict with a lot of our important wildlife habitats, and we are working together with other conservation partners, including folks at Georgia Power, to provide guidance about where to site these installations," she said. "We want to make sure we can have sustainable energy without wiping out important lands for wildlife."

Given her own journey, I asked Erin what parents can do to offer their children the same kind of joyful experience in nature that led her to her present career.

"I developed a more hands-off approach to animals as I grew older," she said. "It's great to nurture kids' fascination with wildlife, but it's also important to remind them that physically interacting with wildlife is not an enjoyable experience for the creatures. It's a balancing act. Due to the presence of wildlife diseases that threaten our amphibian populations today, handling wildlife is discouraged, but you can often learn even more by quietly observing wildlife in their natural habitats," she said.

"If kids are trying to go out in the backyard, get them a headlamp and get yourself a headlamp. Raccoons, frogs, spiders, deer—almost everything in your yard—will have a nice eyeshine, and you can see it with a headlamp much better." Erin said this idea is something she goes over with

FrogWatch initiates, "because a lot of them have never worn a headlamp, or they haven't worn it at exactly the right angle so that you can see eyeshine."

According to the Tennessee naturalist and writer Jennie Moore Ivey, eyeshine is "a reflection caused when light enters an animal's eye, passes through the retina, strikes a reflective membrane, and bounces back. Eyeshine is common among mammals that are nocturnal hunters and foragers," Ivey writes. She found the phenomenon in her own backyard after her hairdresser told her about using a headlamp to see wolf spiders, which appear like tiny, glittering green lights at the edge of the woods.

"Wolf spiders are some of my favorite creatures to show people with a headlamp because you can often find a female wolf spider," Erin said. "They are unusual in that they carry their baby spiderlings on their backs. You can look out and catch hundreds of eyeshines sparkling back at you like diamonds."

Other animals—including dogs and cats—have eyeshine that varies in color. The presence of the reflective membrane also increases their night vision. Humans do not have this membrane, known as the tapetum lucidum, which is Latin for "tapestry of light," but if we put on a headlamp, Erin says, the light shining just above our own eye level has a greater chance of showing us the eyes of what might be out there before us in the dark. Alligator and raccoon eyes glow red. Coyote eyes glow golden yellow. Fox eyes glow green. Otter eyes glow pale amber.

"It's a special experience to have the light reflecting straight back to your eye in a different way than if you just shined a flashlight without having it near your eyes," Erin added. "The spadefoot toad has one of my favorite eyeshines because, for some reason, they shine kind of magenta. I can tell from forty feet away if it's a spadefoot toad, or a regular toad." Erin is still learning from those spadefoots in the dark around her cottage and finding delight in their abundance. "Yeah, they're awesome," she said.

At home, after three relentless weeks of ninety-plus-degree heat in August, I rose at daylight, before the day boiled, to clean out some Bermuda grass that had gotten in my flower bed in Carrboro. I noticed what I thought at first was a wet cluster of leaves sticking to the downspout of the gutter near the front door. I looked closer. It was a little frog with those charming three-toed feet and a pattern of gray speckled skin. I pulled weeds and then watered the bed. When the morning light got a bit stronger, the frog moved to the siding of the house near a spider web that just might yield a meal, I suspected.

I took a photo of my early-morning companion to send to Erin. The night before, I had noticed, she was in Costa Rica posting images on Facebook of two-toned leeches and a little bat drinking from a hummingbird feeder. (Knowing Erin, I figured this was probably her summer vacation.)

Back inside at my desk, I looked up a picture of a Cope's gray tree frog, the species she'd identified singing in the retention pond back in May. A match! I might never have noticed the creature before.

Sadly, frogs and other amphibians are now considered the most vulnerable to extinction among vertebrate animals, according to a long-term study published in the journal *Nature* near the end of 2023. As the international environmental newsletter *Mongabay* explained, more than a thousand experts worked on the study that analyzed some 8,000 species of amphibians. "Between 2004 and 2022, more than 300 species were brought very close to extinction, and 30 percent of these cases were caused by the climate crisis," they reported. Rivers and streams are getting warmer. Forest fires, drought, and hurricanes now move in so quickly, amphibians cannot escape. "Amphibians are disappearing faster than we can study them," said one of the lead researchers.

From now on, I will listen for frog songs with a new awareness of their precarity and diversity. I also now know about the habitat health their music signifies. Our efforts to reverse climate change can't come fast enough.

# 7

# Eastern Screech-Owls

*All night each reedy whinny*
*from a bird no bigger than a heart*
*flies out of a tall black pine*
*and, in a breath, is taken away*
*by the stars. Yet, with small hope*
*from the center of darkness*
*it calls out again and again.*
—Ted Kooser, "Screech Owl"

By the third week of May, the chance of a dusting of snow is remote in the Blue Grass Valley of Highland County, Virginia. But the purely starlit nights still require an extra blanket on the bed, and morning fog is not so quick to dissolve. New leaves on the abundant black walnut trees are not filled out yet, but the lambs, newborn in February, have lost their bandy legs and now romp in the greening fields, racing behind their elders.

The sight of new lambs is irresistible, but we are here for a few days to learn about another group of valley residents, the eastern screech-owls (*Megascops asio*). Only six to ten inches tall, these agile night hunters and furtive fliers are heard and sometimes seen at dusk. Our local guides on this adventure are John and Nancy Spahr. John, a lifetime birder, says that he finds screech-owls "endearing and awesome." He believes that people are often drawn to owls because, unlike most birds, their eyes are part of their humanoid face, staring straight at us with intensity. Their nocturnal habits also make them mysterious.

The Spahrs spend their summers here in Highland County, in the tiny village of New Hampden, on Virginia State Route 640 — a narrow lane with scarce traffic, also known as the Blue Grass Valley Road. With the highest mean elevation of any county east of the Mississippi, Highland County is ranked in the top third in land area in Virginia but comes in last as the province with the smallest population — 2,258 residents, about five people per square mile. No surprise, then, that the environment here seems unusually pristine. The county is divided nearly equally between woodland and pastureland, and the livestock apparently do most of the mowing. Pesticides are used sparingly, we are told. Hardy farmers raise hay, cattle, sheep, lambs, turkeys, and goats. The best-known product from Highland County, however, is maple syrup.

The natural splendor of the Blue Grass Valley is vast, interrupted only by a smattering of communities with mostly historic buildings. We are staying in a cottage on a stretch of Route 640 that runs southwest from unincorporated New Hampden, established in 1858, to the village of Blue Grass, first known as Crabbottom and officially renamed in 1945. Many Scots-Irish families have been in residence here for generations.

Old houses, fronted by short porches, are scattered along the road between the two hamlets, some still bearing the original, never-painted lap siding. Not all are occupied. Several wooden storefronts also sit silent, serving now as storage buildings. Commercial activity tends to cluster around Blue Grass (population 450), which has a library, post office, locally owned bank, hair salon, and mercantile store carrying limited hardware and groceries.

Around 2007, a private developer initiated a plan to install nineteen electricity-generating wind turbines, each to be 400 feet tall and topped with strobing red night lights. The project, to be located on a windy ridge midway between New Hampden and Blue Grass, would have generated clean energy for more than three times the population of the county, but the viewshed in the valley and the safety of birds would have been forever altered. The local debate burned hot and even made the *New York Times* before the developers abandoned the effort.

Economic growth has been forestalled in the Blue Grass Valley in favor of preserving its rural character, much to the relief of some. For others, making ends meet is an ongoing challenge. Most jobs in Highland County fall under the categories of public administration, health care and social assistance, and construction. Poverty in the community, you could say, helps to maintain the environmental riches of the landscape — a complicated choice in so many rural areas of the South and often a cause for resentment among those who experience such stubborn rural poverty.

John and Nancy Spahr live in what was once an Amoco gas station with an attached service bay and small store. The building, dating from the 1940s, had been converted into a residence but was unoccupied when the Spahrs bought and renovated it beginning in 2007. John is a retired physician who specialized in pathology. Nancy is an artist and once owned a gallery in Staunton, Virginia, where they spend the winter months.

John designed and built Nancy's open-air studio, perched above the rocky shallows of the South Branch of the Potomac River, which runs parallel to Route 640. Fifty yards downriver is New Hampden's historic grist mill, which operated from 1816 to 1944. The mill house has also been repurposed as a private family residence with four floors. The waterfall that once drove the mill still splashes over an impoundment shaded by a massive black willow beside a wide and sturdy bridge that leads to the residence.

In recent years, Nancy has been creating mixed-media renderings of birds' nests in her studio. "My bird nest paintings and assemblages celebrate the cycle of avian parenthood—from constructing the nest, egg laying, incubating, hatching, feeding, and fledging the young," she explains. "This cycle is condensed into only four to seven weeks. The last part of the cycle is the empty nest." Nancy works with tar, sand, paint and found items —natural and human made.

John has been an ornithologist by avocation since his teens and still makes several birding trips to exotic locations each year. In 2013, he undertook a one-man census of eastern screech-owls throughout Highland County. He says there may be upwards of ten of these small creatures keeping company in the neighborhood where we are staying, so our chances of hearing and seeing them at nightfall are good.

John meets us on the front porch of his house on a brilliant Monday morning. Donna and I hope to get an overview of his research on owls in the county. He's a genial man, now in his seventies—fit, tanned, and energetic.

"I was always interested in all kinds of birds," he tells us. John was raised in Lancaster, Pennsylvania, the child of displaced persons at the end of World War II who emigrated to the United States from a refugee camp along the Danube in Linz, Austria.

"I started as a kid in Europe, intrigued by wildflowers and always wanting to be outdoors," he says. "Then, when we came to the United States, I spent time with a neighbor boy who knew the local birds, plants, snakes, and salamanders." That neighbor boy was Steve Rannels, a friend since second grade and now a traveling partner on many of John's birding expeditions. (Rannels, a retired physiologist, moth expert, conservationist, and photographer, still lives in Pennsylvania. He and his wife, Sharon, would visit the next day. Steve's advice on the best way to attract moths is considered in chapter 11.)

After his thirty-year career as a pathologist, much of it doing lab research, John says he was ready to do some kind of quasi-academic study on a single bird species. He's a longtime member of the Augusta Bird Club, in the greater Waynesboro/Staunton area of the Shenandoah Valley. He's taught birding courses through the University of Virginia's Osher Lifelong Learning Institute and is a popular speaker with conservation and birding clubs in the region.

Using a standard protocol for surveying nocturnal birds, John meticulously plotted twenty-eight separate driving routes on public roads in the county, each covering ten miles. He would stop each mile along these routes beginning at dusk to listen for screech-owls for two minutes in silence. If no owl was detected, he would make a note and then set up an Apple iPod connected wirelessly to a small, omnidirectional, three-watt speaker on the hood of his truck. He played a prerecorded, thirty-second track of an eastern screech-owl performing its standard calls. The first call, known as a whinny (not unlike a horse's), played for twenty seconds, followed by the second call, known as the tremolo, a purring trill which lasts about ten seconds.

John consistently noted the type of vocal response his recording received, if any, and made an estimation of the distance and compass direction of the responding owls. He noted excess noise and passing traffic in the area. He then used an LED flashlight to see if he could spot the owls he heard. Sometimes Steve Rannels joined him and was able to photograph the creatures.

As weather allowed, between March 2014 and January 2015, John drove and collected data on all twenty-eight routes, seven of them multiple times. Ultimately, he covered 90 percent of the total road mileage in Highland County, making 278 survey stops. He detected at least one owl at 190 of the stops. He found owls on every route except one, which was at the highest elevation.

Based on several categories of habitat he also documented on his stops, John found the largest numbers of eastern screech-owls were hanging out in small clusters of mostly hardwoods adjacent to open meadows, fields, or pastures.

He discovered that in Highland County screech-owls tend to prefer elevations of 2,000 to 3,000 feet, which also happen to be where many wind-damaged trees are found—particularly black locust and sugar maples, which provide natural nesting cavities.

Citing the works of other researchers, Spahr noted the screech-owl's preference to "perch and pounce" for its meals. Unlike barn owls that hunt in pitch dark using their excellent hearing, screech-owls are more visual hunters. His study confirmed that the eastern screech-owl does not shy from living and hunting in proximity to people.

John says that the active sheep and cattle farms in Highland County help to draw an abundance of small mammals that fit the screech-owl diet, while the presence of shallow streams and ponds in the region also provide a complement of aquatic prey. Screechers have a "catholic diet" as he puts it, including beetles, moths, crickets, spiders, salamanders, crayfish, snakes, lizards, earthworms, mice, voles, chipmunks, squirrels, rabbits, small songbirds, and larger birds including blue jays, cardinals, mourning doves, and ruffed grouse. The owls swallow feathers, bones, teeth, claws, and all. Their gizzards give them the capacity to compact this indigestible matter, which is regurgitated as an "owl pellet."

Because of the low population density of Highland County, human interference seldom spoiled John's remarkable survey. Only the harsh winds and temperatures became a challenge. "I did meet all the local sheriffs," he says, grinning. "They all got to know my car. It was easy for me to stop in the middle of the road at night. Not much traffic."

Spahr published his findings in *The Raven: The Journal of the Virginia Society of Ornithology*. He gave his data to a University of Florida graduate student and has continued to monitor a few of his survey routes in the Blue Grass Valley, visiting the stops once a month over the past eight years, to see if he can get a feel for the owl population's stability over time. John also monitors the nesting boxes designed specifically for owls that have been put up by bird enthusiasts across the county.

Patti Reum is a former teacher who runs the guest house where we are staying. She tends a rambling garden and handsome yard, only a few doors down from the Spahrs. Her vintage 1900s rental cottage is popular with visiting birders because it's so close to the Virginia Birding and Wildlife Trail. The cottage is next door to her own house and backs up to a steep hill, where an outspoken donkey comes to the fence several times a day to call out to all his neighbors.

In recent years, Patti, with help from the Spahrs and other volunteers, has dedicated a significant portion of her time to installing and monitoring sixty-nine kestrel boxes in the county. The national population of the American kestrel, the smallest falcon on the continent, has diminished by half since the 1960s and is listed as threatened in Florida. A statewide conservation project in Virginia, begun in 2016, has installed more than 500 nesting boxes across the commonwealth to help secure the population here. Kestrels, like the eastern screech-owl and the eastern bluebird, will often use available manmade housing in lieu of tree cavities and snags to raise their broods of birds.

That afternoon, we set out for the widest expanse of the Blue Grass Valley, where Patti and the Spahrs will be checking a half-dozen boxes. They will count, band, and assess the health of new kestrel chicks. John has promised us a chance to see a brood of eastern screech-owl chicks in a box he put up for them. The baby owls will also be banded.

Our route that afternoon is the Blue Grass Valley Road (State Route 640) beyond New Hampden, heading toward the county seat of Monterey. This road is known as a birding haven and is one of the most scenic drives in Virginia. The writer Oren Frederic Morton (1857–1926), who documented the natural and cultural history of the Blue Grass Valley in 1911, described this landform as "canoe shaped, quite long in comparison with its breadth. It's length, in fact, is that of the county." Morton marveled at how wide open the landscape seemed, "almost suggestive of a Western prairie," he wrote.

We ride with Patti in her SUV. John and Nancy follow in their new electric truck. Once past the Spahrs' house and Nancy's studio, the road curves and ascends. The view opens dramatically. We find ourselves several hundred feet above the bottom of the valley. The road is carved out and curves on the precipitous edge of a ridge. The fields stretch below us for miles on one side of the car. It's as if we are flying. At eye level we follow alongside a red-headed woodpecker soaring in the same direction as the car. Then, below us, a committee of vultures floats toward the mountains across the valley. The farms are laid out in grids, the houses quite distant from one another, with barns and other outbuildings in clusters. Narrow dirt lanes cross the farms perpendicular to us. On the opposite side of the car, we pass dense woods running uphill. The hardwood and evergreen canopy is mixed—by turns maple, oak, locust, white pine, and spruce. Houses nestled above us on the ridge overlook the valley, some set back in close woods. It is all breathtaking.

It seems little has changed from Morton's time. Of the geological treasures in the mountains on the far side of the valley, he wrote: "Their summits arrest the clouds and increase the rainfall within their spheres of influence. The wear and tear of their slopes renews the fertility of the lands below, while within the rocky framework is usually a store of mineral wealth." Native peoples reportedly kept a storehouse of flint for toolmaking in this vicinity, but the embedded limestone, shale, and coal in the valley have not been disturbed by mining, Morton explained.

Patti slows down, and we pull into the driveway of a big, white, abandoned farmhouse, set above the road. The yard is full of blooming buttercups and uncut grass. A robin is barking at us as John pulls in beside a towering black walnut, where he unloads an extension ladder. He sets it against the tree where a nesting box is mounted some fifteen feet above our heads.

John is wearing Crocs, those rubberized shoes favored by hospital workers, which must work well for ladder climbing. He pulls a white bucket from the truck bed and puts a battery powered drill in it. Meanwhile, Nancy and Patti lay out the banding gear, which is stored in a fishing tackle box. They have pliers, a banding tool, and a hanging scale, which will measure the chicks in grams. The metal bird bands are numbered and must be kept in consecutive order. Patti maintains a sheaf of data charts, which she will fill in for each chick along with date, time, and any other relevant comments. The bird's weight is the strongest indicator of health. With experience, John has determined that a small Pringles potato chip can is the perfect size to hold a kestrel chick while it is being weighed.

John climbs the ladder and uses the drill to unscrew and remove a side panel from the bird box. He sets the drill and panel on the flat top of the box and reaches in. There are five chicks, which he places gently one by one in the bottom of the bucket. They are covered in wood shavings, which are part of the nesting material that has been provided in the boxes. John climbs down the ladder and looks up. The likely father of these kestrels, he says, is watching this procedure from a top limb on a tree far up the ridge.

The first fluffy white chick John removes from the bucket is a female, its legs pumping. Its beak is bloody from a recent meal delivered by one of the parents, John says. The presence of some brown tufts among the white fluff is how he knows it's a female. He puts her gently in the Pringles can and records her weight. Nancy slips the band over a leg, and John closes the loop with a special tool. The band is loose, to allow for the chick's growth. Patti is writing the data in her log. Soon, the chick goes back in the bucket, but not before pushing sharp little talons into John's finger. He winces.

The second chick out of the bucket is male. The team methodically goes through the routine with the others until they get to the runt, which weighs in at eighty-five grams. John climbs back up the ladder and clears out some of the baby bird refuse from the box and puts them all back, one at a time. He secures the side panel of the box and descends. "They are pretty site faithful," John says of the parents. "They will come back to nest here year after year, if they can."

Patti notes that not all the boxes that were put up host kestrels. "There were bees in the box across the road," she says, "and now starlings are in there."

The timing of the banding process is critical. It needs to take place before the birds get too big. From their hatch date, it's generally twenty-eight days to the time when they will fledge. These chicks are about twenty days old. Patti says she is grateful to the landowners in the county who allow this process to take place on their property. They report to her on their observations when the parents are getting ready to nest, and again later, when feeding by both parents commences.

At a second site, the team repeats the process, this time being observed through a fence by a gang of black-faced cows with red tags in their ears. Another clutch of five chicks is in the second box, which is mounted in the yard of a small brick church. The kestrel brood seems robust and healthy.

At the third kestrel box we visit, on a fence line down low in the valley, the four chicks are less than ten days old, John says, and they are smaller. Patti explains that as far as she knows, none of the kestrels here have left the valley. But there is competition for the boxes. Starlings sometimes take over a box before the kestrels are ready to nest. Fox squirrels, gray squirrels, flying squirrels, flickers, and even opossums may get into the boxes.

Our next stop is the screech-owl box on this route. It's situated on a steep bank in thick woods, below a house where the owners keep watch on the comings and goings. They stow a wooden ladder near a woodpile, so John doesn't have to carry his own through the forest.

The owl box is considerably bigger than the kestrel boxes, and, as John finds, the mother is still in the box with her brood. He removes the large side panel and brings her out first. She is barely more than a handful. He climbs a few rungs down the ladder so that we can see her. John fluffs her ear tufts. "Owls have feathers all the way down to their toes," he says. The mother's gaze is intense, her eyes a greenish yellow. John loosens his grip, and away she flies deeper into the trees. I worry out loud that she might not return.

*In Highland County, Virginia, an eastern screech-owl mother is temporarily separated from her four hatchlings, who will be weighed and banded before being returned to their nesting box set high on a pine tree.*

"They are so committed to raising their young and passing along their genes," John says. "She will be back."

He fills the bucket with four chicks, about two weeks old. John explains that you can't really tell the sex of an owl reliably except when a female is brooding and has a bare patch on her belly. "But you can tell if these babies are gray or red morphs," he says. The eastern screech-owls at this latitude are mostly gray or mostly reddish brown. John identifies one chick as a red morph, but his experience is showing him something I can't quite see in these mottled gray fluff balls, with their bright, piercing eyes and downward-curving beaks. To me, they all look alike.

To weigh the owls, which are a good bit larger than the kestrel chicks, John has brought brown paper lunch bags to hold them, and one at a time the bag is clipped onto the hanging scale. The first chick weighs 135 grams. The second weighs in at 150 and pees on John's hand. The third is the red morph. It grabs his pinkie mercilessly with its talons. John makes a pained face. The last, the runt, has intensely yellow eyes and is making a clicking sound as it goes through the checkup. Then, back up the ladder they all go, from bucket to nesting box. John secures the side panel. We are grateful for this unusual opportunity.

Some hours later, as dusk comes on, I stand in front of Patti's cottage, right on the edge of the pavement. I can hear the South Branch of the Potomac rushing over stones behind the churchyard across the street. The historic Good Shepherd Episcopal Church is a charming stick-built sanctuary painted white with red doors and tall lancet windows with black shutters on each side of the structure. A massive maple with twin trunks stands guard in the gravel parking lot.

As the afternoon fades and the sunlight angles across the pavement, something in the air is coming down the road, all in one direction. I blink and try to focus. It is a parade of flying insects, with slender black-and-white bodies no longer than an inch, translucent wings, and long, needle-like tails.

These are mayflies, I eventually realize, *Ephemeroptera*, and they have apparently just hatched and are streaming by as if this were a national holiday. They live only a few hours, I've read. I stand there stunned, watching them fly by—scores of them, like a parade of miniature angels weaving in the breeze but hell-bent on following in one direction, straight down the road toward Blue Grass.

Science has proved that these insects do not thrive when freshwater streams and ponds are polluted. When they appear, they are an indicator that healthy freshwater systems are nearby. Check that. We have seen this gorgeous valley in full today.

Mayflies in the larval stage contribute to the health of streams, acting as underwater cleaning systems. They also help in the cycling of nutrients in fresh water and are a significant contributor to the diets of fish. Mayflies have long been studied, and fly-fishers around the world imitate their shapes and behavior when they design and use their handmade flies. Festivals worldwide commemorate their annual appearance.

According to a 2019 article in the scientific journal *Insects*, "Mayflies have been used to illustrate the fleeting and fragile nature of life in literature across cultures and throughout the ages. The earliest reference to mayflies in a written text can be found in the Epic of Gilgamesh, which dates from the 18th Century BC. Mayflies and their short adult stage piqued the interest of early scientific writers such as Aristotle and Pliny the Elder." Mayflies also happen to have the highest protein content of any edible insect, and in some cultures, they are a revered major food source for humans, the article said.

None of this information was familiar to me until I looked it up. But after witnessing the flying insect parade down the middle of Route 640, I was not surprised. It was a sight to behold and an added gift on this outstanding day in the unspoiled Blue Grass Valley.

As the light fades, turning to orange in the distance, here comes John Spahr, walking down the road. He has come with his iPhone and a small portable speaker. He says we are going to call up some owls in the churchyard. Donna joins us from Patti's front porch. We cross the road together. There has been no traffic all this time.

Robins are all around us, making their pipping sounds and hopping around with that side-eyed arrogance they always seem to project. John sets up his wireless speaker on a fence post on the far side of the churchyard, nearer the rushing river. He then straps a small headlamp to his forehead. He does not turn it on. We stand under the massive maple tree. John calls us to be silent and says the owls are likely around. Apparently, they like being near the water this time of day as they hunt for supper.

He reminds us that the whinny sound is a territorial warning to strangers and the tremolo is the way an owl speaks to its mate or future mate. John then plays the recording he used in his owl surveys: first the whinny and then the tremolo. We stand still and listen. Immediately, the songbirds around us—robins, juncos, and chickadees—fuss, sending out warnings that an owl is in the neighborhood because they have heard the speaker. But then we hear an owl answer, too, just like that. Something flies into the maple over our heads, and John turns on his headlamp to find it among the thicket of leaves above us. We get a short glimpse of the screech-owl. We are grinning with delight.

Using recorded bird calls to summon a species is illegal in some parks and can cause confusion about territory among birds, so summoning owls this way is to be conducted in a limited fashion, John says. "It's easy to chase a bird off its territory, so playing a call over and over with a cell phone, particularly, can be bad. Once I get a response, I stop," he says.

We stand there a few minutes and then collect the speaker and walk on down the road, admiring the changing light and the rising mist over the river. It's not even a quarter mile to John and Nancy's house. At the curve leading up and into the valley where we traveled earlier today, a thick stand of trees edges John's property. He sets up the speaker on another fence post and plays the recording toward those trees. Again, there is movement, a rustling in the leaves. We hear an answer, a louder tremolo, and John shines his headlamp into the branches to give us another glimpse of a feathered hunter. He is an expert at spotting them.

The moon is a fingernail sliver, midway up the horizon and over the valley. "When there's a full moon, you can see an owl fly in," John says. He tells us he has been hit twice by owls homing in on him. He's also had multiple birds land very close by here in the months between May and September. We hear the owl call back again, and now the tremolo is fainter in the maples. They are there. That's enough after this fine day. We do not wish to disrupt their evening. It's time for our supper, too.

If you are interested in getting close to owls in your home territory, state and federal parks in the South offer many guided opportunities. Ranger-led "owl prowls" are a program generally offered at the Great Dismal Swamp in both Virginia and North Carolina; at the Audubon Center and Sanctuary at Frances Beidler Forest in South Carolina; at Ijams Nature Center in Knoxville, Tennessee; and many other locations.

# 8

## Dismalites

*Every moment of light and dark is a miracle.*
—Walt Whitman

In the southeastern United States, we're not likely to see the celestial phenomenon known as the northern lights (aurora borealis), caused by electrically charged particles zipping into Earth's atmosphere. The light show we *do* have in the South, however, is a bounty of bioluminescence. We can witness a wealth of living organisms that naturally emit light—insects, fungi, and plants—and most are relatively easy to see.

Lightning bugs, as I grew up calling them, are the most common example of bioluminescence (and the subject of the next chapter). Another form of organic light, sometimes called fox fire, will-o'-the-wisp, or fairy fire, is also found in southern hardwood forests. This eerie, blue-green light is discharged by fungi that feed on decaying wood. According to Kim Coder, a forester with the Warnell School of Forestry, at the University of Georgia, "The best way to see foxfire is in old, moist oak woods where plenty of big dead limbs and old stumps litter the ground." Coder recommends looking for it on a moonless night, far from artificial light. To see the dim shimmer, it's best to navigate the woods without a flashlight. You may also find some wild mushroom species in our southern forests that glow and some millipedes, roaches, and scorpions that light up when exposed to ultraviolet light. These latter insects are telegraphing a message to other creatures that they are too toxic to bother.

Bioluminescence is also present on the Atlantic shoreline. I saw it for the first time walking the beach at dusk on Bald Head, a small island accessible only by ferry off the North Carolina coast. The waves, which were gentle that evening, seemed to be sparkling as they flattened into surf. As I waded into the shallows, my feet and legs stirred up a colorful glitter in the water. Bioluminescent phytoplankton occurs frequently in some places, depending on conditions.

Vendors on Florida's space coast offer guided kayak tours at night to observe sea sparkle, but in this case, all that glows is not good. Unfortunately, a toxic magnification of this phenomenon is on the rise. Sea sparkle is created when phytoplankton is exposed to an overabundance of nutrients from fertilizer, wastewater, and stormwater runoff. Add abundant sunshine to warm the water, and sea sparkle may develop into an algae bloom that, at its extreme, produces red tide, a pollutant dangerous for humans to swim in.

Creatures known as Dismalites, the subject of this chapter, thrive in an obscure location called Dismals Canyon, in Alabama. These grain-of-rice-sized gnat larvae live in damp moss and emit pinpoints of light. They have been described as "glowworms" by the owners of the canyon, and the site has been drawing small groups of curious observers for decades.

The species *Orfelia fultoni* was discovered in 1939 by the entomologist Bentley Ball Fulton (1899–1960). He found the glowworms along a hillside spring behind the house he was using as a summer field laboratory in Glenville, a mountain village in Western North Carolina near the historic resort town of Cashiers. Fulton had been studying much larger insects, but the faint emissions of light coming from the bank were most unusual. He collected specimens to take home to Raleigh.

The abstract of the scientific paper that Fulton eventually wrote, announcing his discovery, is lyrical, almost a poem: "Points of bluish light glowing at night on the damp earth about a spring in western North Carolina led to the discovery of luminous, lochetic larvae living in webs after the manner of spiders. The worm-like larvae remained concealed in crevices during the day but at night they crawled out on the webs and exposed their glowing bodies." ("Lochetic" means "lying in wait for prey.")

Fulton had no appropriate reference books on hand, so he couldn't determine the possible species. His primary research was with katydids—those raucous, grasshopper-like insects that sing with their wings at night in late summer and fall. Fulton tried to coax the newfound larvae to reach maturity so that he could determine their kind. They seemed to behave like spiders.

He wrote: "When I confined the larvae in vials or Petri dishes on beds of moist peat, they built webs and lived normally while they were kept in a cool environment. Unfortunately, the field work was discontinued before any of the larvae matured. Two lots of larvae removed to Raleigh in June and July died from exposure to heat, and no adults were reared during the first year. A year later I took larvae to Raleigh in an iced container and upon arrival kept them in a cool basement."

From that last group of living larvae Fulton managed to produce several adult insects. A colleague in Raleigh identified them as a yet-undescribed species of *Platyura*—a genus of fungus gnats.

In 1941, the year Fulton published his findings, the Nantahala Power and Light Company completed a dam on the west fork of the Tuckasegee River and flooded the surrounding valley. Lake Glenville, with a depth of 600 feet in places, covered the original town of Glenville and became the highest reservoir ever built east of the Mississippi River. The lake presumably buried the hillside and spring where Fulton first found the gnat larvae. These creatures would carry his name, however, and Fulton would spend the rest of his distinguished career in the South, teaching entomology at North Carolina State College from 1928 to his retirement in 1954. He was elected to the North Carolina Academy of Science in 1955.

It has since been proven that these North American "glowworms" evolved independently from their more famous counterparts in Australia (*Arachnocampa flava*) and New Zealand (*Arachnocampa luminosa*), where they are easily seen by touring groups because they thrive in dark caves that have been made accessible to visitors.

The glow can be more difficult to spot at night in remote Appalachian forests, where the larvae are embedded along the moist banks of rocky streams—"almost microscopic" as the Knoxville, Tennessee, firefly expert Lynn Frierson Faust put it. Nevertheless, this distinctive North American fungal gnat has been documented in many more sites across the southern Appalachians, including Anna Ruby Falls in Georgia; Hazard Cave and Pickett State Park in Tennessee; in Dry Falls, Highlands, Grandfather Mountain, and Mount Pisgah in North Carolina; and in Shenandoah National Park in Virginia.

By my reckoning, Alabama's Dismals Canyon and Georgia's Anna Ruby Falls have been the most publicized sites of this marvel. For more than thirty years, Georgia State Park rangers have been leading night hikes at Anna Ruby Falls to see the larvae, which show up in May and June. They are best viewed on the return hike from the falls. Alabama's Dismals Canyon is open to visitors March through November, though guided night tours to see the glowworms may be scheduled only on weekends. Peak times to see the Dismalites are late April through May and again in late September to October, according to the website. Predicting when the glowworms will show up in their greatest numbers at any site is a challenge, dependent on seasonal weather and moisture levels.

I was drawn to the Alabama destination because the proprietors present the attraction with an old-fashioned dose of showmanship as a magnificent oddity of nature. The allure is not unlike Rock City and Ruby Falls, near Chattanooga, both of which are commercial attractions, made compelling by decades of advertising on barns and bird houses in the rural South. I never forgot the fun of these weekend destinations in my childhood, and even as a second grader, I could recognize the hype. To this day, Ruby Falls involves an elevator ride deep down into a rock cavern, where a tour guide walks you through a sequence of garishly lit stalactites and stalagmites. Each tour ends in pitch dark at a destination where a waterfall can be heard splattering nearby. Then, with great fanfare and the flick of a switch, Ruby Falls is illuminated with kitschy red lights.

Though the Dismals Canyon website has recently been modified to present a more science-based approach to what is essentially a natural spectacle, the use of superlatives about the experience is over the top. And would I go there again? Yes.

Part of the Appalachian chain, Dismals Canyon is a few miles south of Phil Campbell, Alabama, a town of fewer than 1,000 residents located thirty miles due south of the legendary recording studio at Muscle Shoals. Lisa Buck, a retired teacher and one of the organizers of Alabama's annual Cahaba Lily Festival, agreed to accompany us to this remote spot in May 2022. She knew the area and taught middle school in nearby Hackleburg for four years. But like many Alabamians, Lisa had never visited the glowing attraction, though she had heard of it.

The entrance to Dismals Canyon is unexpectedly forbidding. Rough-hewn cairns of gray stone secure a sign that declares the destination a National Natural Landmark. Nearby, stone pillars with faux torches flank a massive gate made from timber posts woven together with thick vines and topped with menacing, wrought iron spear points. A bronze casting of snarling lions face off at the midpoint where the gate opens. The stone wall that continues beyond the gate is set off by lanterns with orange bulbs flickering day and night.

Among the old-growth trees just inside the property are several rare, big-leaf magnolias (*Magnolia macrophylla*). They are dinosaur-scale, with leaves three feet long and creamy blooms larger than a human head. This privately owned eighty-five-acre canyon has been known as a place for recreation going back to the 1920s. Indigenous peoples also occupied the canyon long before European settlers arrived. The massive sandstone formations where the glowworms live date back to the Paleozoic era. The site's designation as a National Natural Landmark is a contemporary honorific, issued in 1975 by the US Department of the Interior as part of an effort to recognize and encourage the conservation of sites that contain "outstanding biological and geological resources."

"We have the last primeval forest east of the Mississippi," the canyon's resident biologist, Britney Slappey-McCaffrey, told *National Geographic* when they came to photograph the glowworms in 2021. "What that means is that our forest has never been touched by ax or fire, so it's all old timber. It's almost impossible to get down there with any kind of machinery for farming or lumber." That the canyon has gone relatively untouched speaks to its remote location, but it also seems to have remained undisturbed in the last decades primarily because of the tiny glowing gnat larvae.

Beyond the parking lot inside the gates, visitors enter the canyon through a general store, where the proprietors sell candy, toys, T-shirts, bottled water, chips, and sundry souvenirs. An old-fashioned soda fountain on one side of the building offers sandwiches and ice cream on some days. Two rental cabins and a camping area are available by reservation.

We paid for our admission, got identification bracelets, and signed release forms taking full liability for any mishaps on the hike ahead of us. The pleasant woman at the ticket booth gave us a map and pointed out the path we should follow. There had been recent rain, she said, and we should watch out for slippery rocks and the possibility of mud on some of the wooden walkways. She noted the swinging bridge on the map and said it was off-limits. We were cautioned not to touch the rock faces, to take nothing from the site, and to leave no litter behind. The map offered more stern warnings in print and emphasized the prospect of fines. The tone was not unlike the forbidding imagery at the entry gate. The woman at the booth pointed us out the back door of the store to a porch at the far end of the building. From there we began our descent.

Wooden stairs led to a paved walkway running downhill and flanked by an iron railing at the edge of a sheer drop-off. Boulders lined the higher side of the path. Eventually, the walkway ended at a stone patio. Here was an empty swimming pool that was apparently a key amenity for visitors at one time. Now the box of concrete was locked behind a metal gate with a sign that read "No Lifeguard on Duty." Arrows directed us to a large wooden deck set atop concrete blocks and surrounded by more wrought iron railings. The deck was not in the best shape. Some areas were patched with plywood. Missing boards were marked with caution tape and traffic cones.

We steered around the hazards and headed directly toward the long, steep stairway that would carry us down another thirty feet to the canyon floor. The stairs, also made of wood, were open so that we could see the rushing water below us. According to the map, the whitewater was coming from Rainbow Falls, upstream, where Dismals Branch first drops into the canyon. This waterway had once powered a grist mill that a flood in the 1950s had destroyed. A natural pool farther upstream had been used for baptisms in the 1800s, the map explained. The railings on the stairs were weathered two-by-fours. Roofing shingles had been hammered on the stair treads to help secure our footing. At a landing halfway down, we stopped and studied the rushing branch. A thick brown snake was sunning on a rock. Nearby, an armadillo was scavenging under an outcropping that we also would soon pass under beyond the stairs.

The stone formations of the canyon were formidable, towering over our heads and festooned with bright green moss. A group of three young people, employees of this enterprise, appeared ahead of us, sweeping a boardwalk and a stone walkway that had been slicked up by rain and mud. This jumble of gigantic rocks, said the map, was the result of ancient earthquakes.

Now on the trail, we came upon a handwritten sign propped on a rock that declared "Caution: Snake Den." Unnerved, we kept moving. Later we would learn that the sign was meant to deter visitors from a vulture's nest behind the rocks where young birds had yet to fledge.

There were sixteen named destinations on the trail map, most of them geologic features with evocative names—Pulpit Rock, Weeping Bluff, Secret Falls. Burr's Hideout, the map said, was used by an Alabama outlaw to evade capture in the 1800s. Weeping Bluff was described as having "the face of an Indian Maiden." The rock face was weeping, the map explained, with "tears shed by the Canyon for the loss of its only true friends, the Chickasaw Indians."

Northwest Alabama was part of the Chickasaw domain for thousands of years. The people who knew these grounds as their sacred home ceded the land to the United States in 1818. Andrew Jackson's army later rounded them up under the Indian Removal Act and reportedly held them in the canyon for two weeks. Then they were forced to march to Muscle Shoals to join other Native prisoners on the Trail of Tears. They arrived in Oklahoma in 1838. The Chickasaw people who survived the journey would maintain their cultural traditions at a new home, now in thirteen counties south of Oklahoma City.

Temple Cave in Dismals Canyon is thought to be a site where, some 10,000 years ago, ancient peoples took shelter, built fires, and ground corn on the rocks. Spear points dating to the era have been discovered around the canyon, but no archeological excavations have been performed here to confirm whether Paleo Indians, who entered the continent some 20,000 years ago, had occupied this area.

In other notes about the geologic formations, the map explained that Girl Scouts often camped in the canyon in the 1920s. They created a pool to collect water and used an ancient Indian site, now called the Kitchen, to build their fires for cookouts.

Donna, Lisa, and I didn't talk much. Deeper in the canyon, the air was heavy with moisture. The wind moved only through the very tops of the old-growth trees that towered above us. We were far below that canopy of green leaves in this labyrinth of wet sandstone. Massive boulders provided natural bridges to walk over. Others narrowed into dark crevasses. Vertical walls—some sixty feet tall—were slick and forbidding. Stone paths had been laid to guide us through the sandy soil. In one spot, crosscuts from a large tree trunk substituted for stepping stones. Roots as thick as tree trunks muscled over boulders flanking Dismal Branch as it rushed through narrow passages of rock. Smaller roots the size of snakes undulated over the rockface moss. Millipedes crawled in and out of view. Trailing arbutus hung down a sheer rockface with tufts of lush ferns running alongside. Younger trees leaned and twisted oddly toward available light while the occasional big-leaf magnolia stood sturdy and unmoving. The smell was earthy and wet.

Two eastern hemlocks (*Tsuga canadensis*) marked the halfway point on the trail where we crossed the rushing branch and headed back on the other side of it. The taller of the two hemlocks had been topped by an ice storm in recent years, the map said. It had once been 139 feet tall, with a fifty-foot spreading crown—a champion tree, meaning the largest of its species in Alabama. We could only lay hands on its nine-foot girth and look up to see the broken snag. The sky was remote, another world above us.

I found the scale of the canyon to be both spooky and spiritual. The sound of dripping water echoed in every direction. The presence of the ancients was palpable. Rocks and roots traversed our path. They made us move mindfully as we picked our way back along a higher bank above the

*In Dismals Canyon, Alabama, thick wet moss covers ancient rock, harboring gnat larvae that glow like stars after nightfall in May.*

water. We were looking down at the route we had already followed on the other side of the branch. When we finally reached the swinging bridge, we could see why it was off limits; the treads were hinged to the sagging rails, and ragged rope flanked the sides. It did not look safe. We glanced into Witches Cavern, an impossibly narrow crevasse winding into blackness beyond. At last, we came to a fixed wooden bridge and found ourselves in sight of the long staircase back up and out of Dismals Canyon. I was trying to imagine how we would navigate these trails again in a few hours, in pitch dark, when we hoped to witness the shimmer of glowworms.

On our way to find an early supper, Lisa told us how this section of northern Alabama had been prone to devastating tornadoes. In 2011, she said, a tornado had come alongside US Highway 43, where we were driving at that very moment. Measuring a five on the scale meteorologists use to quantify strength, the twister destroyed three-quarters of Hackleburg, a town larger than Phil Campbell by a few hundred residents. The school where Lisa taught had to be completely rebuilt. The tornado also flattened a Piggly Wiggly grocery and the Wrangler Jeans plant in Hackleburg, eventually dropping new pairs of jeans from the sky, some forty miles away, in Courtland, Alabama. As the storm headed northeast, its path was a mile wide, news reports said. It must have come precariously close to Dismals Canyon. Then, without losing any strength, it ripped into the town of Phil Campbell. That storm, on April 27, 2011, was part of the largest tornado outbreak in US history, and the so-called Hackleburg–Phil Campbell tornado—one of the many that were spawned that day—was the deadliest in Alabama history, with seventy-two deaths and peak winds of 210 mph. That was an aspect of the wild South I will always prefer to miss.

We drove through Hackleburg with its impressive new school buildings and shiny mobile homes. We then backtracked all the way to Muscle Shoals, where we found a barbecue joint for supper. Soon it was time to return to the canyon and wait for sunset.

Our seasoned canyon guide, Bill Ogle, met us outside the gate and introduced himself as a fabricator for a manufactured home company by day. By night he met brave hikers at the locked gates of Dismals Canyon at dusk, eerie lights flickering, and let them into the park. The front pockets of his blue jeans were bulging with miniature flashlights that emitted red light, which he said would help to preserve our night vision and would not register with the Dismalites. Bill wore well-seasoned hiking boots and a warm shirt. He pointed out our designated parking spaces, and then we headed for the porch of the general store.

There were eight of us counting Bill. Two couples were already on the deck of the store waiting, the Services from Birmingham and the Johnsons from Montgomery. They were staying the night, one couple each in the two rental cabins. Donna, Lisa, and I introduced ourselves, and we made small talk until Bill was ready for us. He asked first if mosquitoes had eaten us alive on our day hike in the canyon. I thought he might be about to offer us bug spray. We all shook our heads, no. We had not been bitten. Had we heard any birds in the canyon? "No," we said again.

"The Dismalites were busy eating the mosquitoes, and the birds stay high up in the tree canopy," Bill said. How had we missed this? Then Bill introduced us to Tick, the resident cat, who wandered at that moment into the circle of human guests on the deck outside the store.

Bill told us we must not touch any rock walls or moss on our canyon hike. The salt from our hands would kill the Dismalites, he said. The canyon walls in one area had been defaced some years back, and no Dismalites had ever come back to that spot. "But the Auburn University entomologist who was here earlier this year told us this is a banner year," Bill said, "with tens of thousands of larvae in residence." He handed out flashlights. "Shall we go?"

We rose from our seats on the upper deck and set off down the paved path to the second deck and the ominous stairs. Darkness was coming fast now. At the head of the stairs, Bill suggested every other person turn on a red flashlight and point it down to navigate the steps. He told us to help each other and to take it slow. He was in the lead.

It was dreamlike going down those long stairs again, this time with minimal light and the now seemingly louder sound of rushing water below. When we reached the bottom, Bill had us turn off our lights and stand for a minute, getting our bearings and adjusting to the darkness. We stood in the deepening evening, listening to the water, and then Bill led us forward. We moved like a caterpillar, in sections, close and connected. It was now a blind kind of dark. Donna and I were just behind Bill, though we could not see him well. Rocks hung over our heads. Then, coming around a curve in the path, we heard Bill hit the dirt with a thump and a groan. He had tripped on a root and careened forward, landing hard. Then I heard him patting himself off, as he ducked under a stone overhang and turned to face us. He laughed and said it happens sometimes. He clicked on his red light and pointed to the roots for the rest of us to avoid.

"I took a face plant by tripping on a root and slamming into a rock on my very first guided tour," Bill said, still catching his breath. "With forty boy scouts." He took a gulp of air. "Now I only take smaller groups," he said.

We headed slowly into the rock passage, and Bill pulled out an ultraviolet flashlight. It shone purple on the rock face to our left. There, a millipede was traveling with all its legs in motion. The creature glowed yellow in the purple light. "Yep, he's poisonous," Bill said. Everyone had a good look as we lumbered on, lights out again.

We came to a rock face that Bill called Indian Head Profile. He stood beside it outlining the features with his blue light—a mohawk haircut, a prominent nose. "I am part Cherokee," Bill said, "and I feel a strong connection to the spirits of the Native people who once lived here." Then he shut off his light and asked us to make a quarter turn to the left and face the wall of rock beside us. He said to extinguish all flashlights and to shut our eyes and count to thirty. It felt odd and intimidating to be standing with a group of strangers in a foreign dark, eyes closed, with creatures crawling on the rocks in mind's eye. I breathed and counted slowly, sensing the others around me doing the same, but feeling also blank and dull, as if ready for sleep.

At Bill's eventual prompt, we opened our eyes, and there they were: tiny pinpoints of light covering the rock wall, top to bottom, and dotting the landscape at our feet. People whispered exclamations. "The ones down below are lightning bug larvae; these up here are the Dismalites," Bill said. We stood frozen, in awe.

"Looking at them straight on is elusive," Bill said. "Use your peripheral vision, if you know what I mean, and you'll see more." We remained quiet. The couples spoke a word or two to each other—expressions of marvel and gratitude. Fear fell away. The lights were tiny, evanescent, but undeniable.

Then I looked up. The night sky above us was clear, star studded. The points of light were continuous from ground to rocks to sky. It felt like we were in an airplane, flying over a big, lit-up city. Then someone noted occasional lightning bugs flashing over our heads in the trees. We waited and watched for them. The effect was uplifting.

Bill encouraged us to move on, but as if in slow motion. Now we were headed into a tighter space between rocks, he said. When we stepped into the entrance to the next formation, he pointed out several constellations of larvae on the walls. "My granddaughter calls them living stars," he said. The group hummed with recognition.

Bill pointed to a constellation he called the "Civil War cannon on wheels." "Do you see it?" he asked. "Use what I call your alien eyes, squint up and look to the right." We moved our heads to see it side-eyed. I moved ahead out of that tight spot toward another up ahead. Why did we always have to see the Civil War?

"And here is what I call the 'redneck water slide,'" Bill said, suddenly turning on his white light. We were approaching a sheer drop-off of loose dirt and mud on one side of the path, ending in roots and water below. "Stay to the side," Bill said, motioning us along with his light. I had no idea where we were in relation to the map we had used on our day hike.

"Now, you are going to have to duck in here," he said, turning off the white light once we had all passed the water hazard. We were back to red light. "Bend forward and put your back to the rocks. Step sideways. It's okay if you rub the back of your shirt on the wall if you have to, but please avoid contact between rock and skin."

We each moved in turn—sideways, one at a time, ducking and contorting our way through the first part of the curving passage. "Watch the roots; step over that rock up there," Bill said. It was a wide rock, requiring a long stretch, one leg at a time. We were fully surrounded by the stone. It felt claustrophobic. "This is the Witches Cavern," Bill said with satisfaction. Now I realized we were taking a much shorter loop in the canyon than we had in daylight. Thousands of Dismalites were here on the walls and at our feet, everywhere, glowing mirage-like, shimmering. Without the night sky above us in this confined space, they seemed more intensely blue. It was a dream. I felt like an astronaut. I tried to imagine what the Native people thought the first time they saw the lights at night.

Bill edged up next to me and pointed to a glowing spot. "The larva is in there, in that moss, no bigger than a grain of rice." He shone his light on the moss and the blue winked off. "See, they quit on you, if they know you are looking."

Slowly, carefully, we moved on.

"Now we are coming to Swinging Bridge," Bill shouted from behind us, over the roar of water ahead. "One person at a time. Wait to step on the bridge until the person in front of you is stepping off." When it was my turn, my legs were wobbly. I felt like I was being jettisoned off the bridge up into the air when one of the men stepped on the bridge behind me. Donna practically flew off, too, when her turn came. But we made it. We had seen something new and unexpected. We climbed the rickety stairs one more time with a feeling of relief and awe.

Tick the cat greeted us on the patched deck, weaving among us, tail up, as cats do. Bill told us the owner of the canyon is a woman in her seventies from Mississippi. He hopes the property will stay private. If he owned it, he said, he would put the mill back in operation to generate power and take the whole place off the grid. "It is important to preserve this place," Bill said earnestly. We thanked him for the tour, holding onto the vision of the glowing points of light that reached from our feet to the stars.

We have pictures only of our daytime hike in Dismals Canyon. Photographing Dismalites requires a long and steady exposure, and even then, I suspect that the photos I've seen, such as those from the experts at *National Geographic*, are enhanced. Donna, Lisa, and I carried the images of the sparkle in our brains, recalling the odd, sideways glances required to take it all in, the meeting of glowing larvae and celestial stars at the edge of the rock and sky, and the yellow fireflies winking in the trees above. No matter how sophisticated our technology, it seems fitting that some spectacles, especially this one, remain beyond capture, and beyond description.

# 9

# Fireflies

*I watched the fireflies as I looked down into the
bottom land. . . . The scientist would explain it
probably as the mating of insects, but how would
he account for the joy it raises in the beholder?*
—Harlan Hubbard

For a child growing up in the rural South in the
past century, lightning bugs seemed like shooting
stars within reach, fireworks at eye level, sudden
birthday candles coming out of pitch dark, and
the surest sign of summer's arrival. For my generation, spotting the first of these luminous insects each year can kindle deep memories of a
specific place and time—the cool, silent magic
of flying sparks that can be caught and held in a
jar, still glowing.

The charismatic presence of fireflies feels like the truest example of the
fabulous ordinary. Yet I never knew the complexity of their mechanisms
of illumination, their range of genus and species, or the variations in their
love language of light. It was time to go to lightning bug school.

The sad news is that the firefly, as the Swedish taxonomist Carl Linnaeus
first called it in 1787, is losing habitat and becoming less ordinary in many
places. Neither bug nor fly but a beetle kin to the ladybug and June bug, the
firefly's existence today is challenged constantly by earthmovers clearing
out forests and fields and draining wetlands. Fireflies are also victims of
leaf blowers that interrupt the natural decomposition of organic matter

in fallow beds, lawns, and along forest edges. Leaf litter is critical to every stage in the lightning bug's life cycle.

In their prime time of reproduction, fireflies must also contend with the bright lights of neighborhoods—rural and urban—and too many big-city bulbs burning till dawn, confusing their flight patterns and mating rituals. Pesticides spread on golf courses and manicured yards kill lightning bugs. Even the mosquito deterrents that humans spray on themselves can kill them.

And yet, along with this wholesale disruption of firefly habitat, the public urge to learn more about these creatures has suddenly quickened. Ever since 1995, when experts confirmed the presence of synchronous lightning bugs (*Photinus carolinus*) in the Great Smoky Mountains National Park in Tennessee, people have been flocking to witness the phenomenon—so many that an annual lottery activates in April, about six weeks before the insects have begun their annual breeding cycle.

Blinking in synchrony is a high level of magic, and the spectacle that has come to be known as the light show in the Smokies has lifted fireflies into the league of performing artists. Like synchronized swimmers, dancers, and ice skaters who put on elegant displays of discipline and rhythm, lightning bug shows have become a coveted ticket in the region. In 2023, more than 20,000 people applied for the 120 tickets issued per night for eight nights in the Smokies.

Science has long known that the firefly's light comes from a reaction between two chemicals: luciferin produces light when it encounters the catalytic substance luciferase. It's the same chemical reaction that excites the Dismalite glowworms in Alabama. Both Latin names come from *Luciferus*, which means "the light bringer," associated in ancient mythology with Venus, the morning star.

Through the dedicated research of a relatively small group of scientists, we also now know that more than 2,000 species of fireflies exist worldwide, though much remains to learn about them. Lightning bugs belong to the insect order Coleoptera (beetles) and the family Lampyridae (meaning "shining fire" in Latin). There are seventy-five species in eastern and central North America—a surprise to many folks I've talked to who, like me, thought a lightning bug was a singular insect.

In the South, the species commonly called big dippers (*Photinus pyralis*) show up before sunset beginning in June. Dippers are probably the most recognized lightning bug in the region. With their lights switched on, these fireflies drop down and then rise halfway back up, writing the letter J on the blackboard of evening before they wink off. Experts describe them as being the coyotes or the robins of the firefly world because they are so widely dispersed across the country, as far west as Texas and South Dakota. They are also more adaptable to urban and suburban habitats.

In daylight photos, big dippers look familiar to me, based on the coloration of the lightning bugs I collected in jars during my childhood in the woods outside Atlanta. Crawling in the hand, they are mostly black, edged by a white line all the way around. The white line also bisects the insect's back where the two wings unfold in flight. The head is orange/red with a black dot in the center and two antennae protruding from the top. Big dippers are found in all kinds of environments, from fields to woods, roadsides to orchards. Because they start up so early in the evening, they seem less affected by light pollution, scientists say. They may even manage to produce two separate generations over a single season. I am sure I have seen big dippers all my life, but many other species are common in the region.

Other firefly species vary in size, habitat, mating period, body type, and their characteristics of illumination: color (green, blue, yellow, amber), frequency of flash, average height in flight, and flight patterns. External influences, such as topography, altitude, and the ambient air temperature and moisture on any given summer evening, can also affect the frequency and speed of flashing. Warm weather speeds up the flashing of some species, and overall, the flash patterns are remarkably varied.

The one behavior that all firefly species share, though, is the fierce drive to reproduce, which gives rise to the range of light shows that one can witness in the wild. Observing the synchronous species, however, would be our first field experience.

Donna and I watched a story on *CBS Sunday Morning* showing the crowds of lightning bug lottery winners being bused by trolley to the viewing area in Elkmont within Great Smoky Mountain National Park. We decided to contact Liz Domingue, a private guide, and the owner of a business she calls "Just Get Outdoors." Headquartered in Maryville, Tennessee, Liz is a professional naturalist who has been leading excursions—daylong and weeklong—around the region and beyond since 1999. During the narrow window of lightning bug luminosity, she'll take small parties to areas around the Smokies, away from the hubbub, to find a synchronized display.

Liz earned a graduate degree in wildlife ecology and conservation biology at the University of Florida and studied wildlife biology as an undergraduate at Cornell University. She says her interest in nature and her love of the outdoors "has been a lifelong pursuit, at least since about age seven." That was when Liz first saw the word "naturalist" and immediately knew that was what she wanted to be in life. Her parents had started her on the path early. Liz's mother wrote in her baby book that her daughter's first word was "bird"—something she heard while being held up to the window to look outside. Today, Liz's excursions include trips to Virginia's Shenandoah Valley, the Florida Everglades, and several destinations beyond the wild South. She has written a field guide on paddling the Okefenokee and is coauthor of *Butterflies and Moths of the Smokies*, published in 2019.

We met Liz, along with about a dozen other eager explorers, at the Townsend Visitors Center, west of the commercial gateways of Gatlinburg and Pigeon Forge. The evening was exquisite, the clouds glowing pink as our caravan drove through the forest toward a quiet trailhead where we could line up our folding chairs and face into mature woods. We could hear water rushing close by.

As we settled in without phone light or flashlight, our eyes adjusted to the dimming light. Liz explained that nineteen species of fireflies have been confirmed in the Smokies. She said the synchronous males we would see are very bright flashers. The females also wink on, albeit more dimly. "They know each other by their flashes," she said.

"Just so you know what to expect," Liz went on, "the males don't flash all at once. They do it in groups, kind of like 'the wave' you see in sports stadiums. They will flash six to eight times and then go dark for six to eight beats. That's the pattern." Two kids in our group raised their arms in sequence, imitating "the wave," and then poked at each other, grinning.

"We used to think synchronous lightning bugs were only here in the Smokies and in Thailand," Liz added, "but now we know there are two other synchronous species in the South."

Liz mentioned the book *Fireflies, Glow-worms, and Lightning Bugs*, the first comprehensive guide to the species, published by the University of Georgia Press in 2017. The author, Lynn Frierson Faust of Knoxville, Tennessee, is widely known as "the Lightning Bug Lady." She's been studying lightning bugs for three decades and has served as a consultant on several nature films and television series.

Faust was the first to draw attention to the synchronized species at Elkmont in the 1990s. She also describes the other two synchronous species in the region. The cattail flash-train firefly (*Photinus consimilus*) lives in wet and swampy areas from Florida to Canada, especially along the eastern coast, and as far west as Missouri and Arkansas. The cattail males flash very quickly, five to seven times as they fly, and the females answer with a double flash from the ground. (Scientists call consecutive flashing "a flash train.")

*Photuris frontalis*, also known as the snappy single synch, is the third regional synchronous species. They range over the Deep South and rapidly flash a deep yellow. They are present during a short season of two to three weeks, which begins earlier the farther south you go—April in Florida, May in Georgia, and so on, up as far as Maryland. The snappies seem to prefer cypress forests, and, as Faust explains, "your chance of observing their synchrony increases as their density increases. Lone or loosely spaced individuals, whose quick flash is easily recognized, simply march to their own rapid beat, unsynchronized." Liz told us that recently, snappy single synchs have been studied in Tennessee, on federal property at the Oak Ridge National Laboratories.

"When you think about it, frogs call to attract mates, and it's the same with fireflies," Liz said. "The males are always competing, trying to be the first to flash so the female will choose them. This is sexual selection at work, so the theory goes: females seem to be choosing males based on who flashes first. Females also need the six beats of darkness to be able to answer the males and for the males to be able to see them."

Doing research with a team of colleagues, Lynn Faust found that "*Photinus carolinus* females will refuse to respond if multiple male flash patterns are presented to them randomly—a process called *jamming*—instead of in synchrony. Inundated with a cacophony of flashes rather than an orderly display, these 'opinionated' females refuse to answer any males until those out-of-synchrony males get their flashing acts together again." So the light show in the Smokies is indeed a complex routine for all performers.

Lightning bugs lay their eggs on the ground, and in 2016, when some 200 wildfires spread across North Carolina, Georgia, Alabama, and Tennessee, the numbers of lightning bugs dropped dramatically. "Not because of the fires," Liz said. "It was because of the drought." The presence of water is key to these insects, especially in the larval stage.

As for predators, Liz explained that lightning bugs don't taste good, which reduces their chance of becoming prey. They carry in their bodies a bitter-tasting chemical called lucibufagin (lucy-BOOF-again)—a cardiac toxin like the organic compounds found in toads and in foxglove and milkweed plants. Birds and bats will not eat fireflies because of the taste and, presumably, the danger of poisoning. I remember the bitter smell on my hands as a child after catching a few lightning bugs and putting them in a jar. When they are threatened, fireflies ooze the lucibufagin, a sticky white liquid. Of course, warding off potential predators is essential when your taillights are always giving you away.

Lightning bugs do fall prey to spiders and their webs, however. Harvestmen, also known as daddy longlegs in the South, often poach from spiders' webs and carry off fireflies for a meal. Despite the defensive chemistry, there are other predators, too. Some forms of bacteria, nematodes, roundworms, mites, and assassin bugs will kill fireflies. One species of fly deposits its eggs inside a firefly's abdomen. The fly larvae, when hatched, will then eat the firefly from the inside out. Some fireflies flash rapidly when in distress, often after being caught in a spider web. They can also continue glowing for an hour after death, Lynn Faust observed.

The sole purpose of adult lightning bugs is to mate, so once they are fully grown, they don't eat. Faust noted that the only exception is when adults may sip a little common milkweed sap to enhance their toxins against predators.

Fireflies in the larval stage, in contrast, are voracious consumers of slugs, snails, earthworms, and soft-bodied insects. They inject their toxins to paralyze and dissolve their prey and then consume them from the inside out. Turnabout is fair play, I guess.

One species of adult firefly, the female of the *Photuris* genus, has been observed faking the female flash pattern made by the genus *Photinus*. When *Photinus* males respond to the imitation, they are promptly consumed by the femme fatales. A team of researchers from the chemistry department at Cornell University proved that the *Photuris* females gain lucibufagins from the meal, which they are unable to produce themselves. The toxins in turn make the femme fatales undesirable to one of their chief predators, a species of jumping spider.

As darkness arrived at our little cul-de-sac in the woods in the Smokies, the light show began. It took a while, but the numbers of fireflies gradually increased, building their themes of light like the first movement of a symphony. They established rhythms, and soon we could discern the synchrony, counting to six during the dark intervals and then following the new waves of light as the males made their best attempts at a chorus line of seduction. The whole on-and-off scheme was spellbinding. As observers, we quietly gave voice to our excitement as the lights were unleashed. The flashes seemed brighter as the evening grew darker.

I was thrilled to see this display in a quiet dome of forest, away from the crowds that threaten to love the Smokies to death. In her book, Lynn Faust expressed her mixed feelings about being the scientist who first brought attention to the synchronous fireflies in the Smokies and how it has developed into such a human spectacle. But we must know what we threaten with our own lights, what creatures live in our old-growth forests at night—their cool, silent magic of flying sparks. (One-fifth of the 500,000 acres in the Smokies are still old-growth forest.)

Seeing synchronous fireflies is a great introduction to the complexity of these insects, and they may be seen in several other public places in the region. Joyce Kilmer National Forest, in extreme Western North Carolina, welcomes firefly seekers, who gather in the main parking lot from mid-May through June carrying the required red or UV-purple flashlights. Observers are encouraged to follow the trail that goes into the forest and to choose a bench along the way as they wait for dark and the lightning bugs that will appear.

The Audubon Center and Sanctuary at Beidler Forest near Haleyville, South Carolina, also hosts occasional night walks on weekends throughout the year. This Lowcountry habitat is the world's largest old-growth cypress-tupelo swamp forest, which might be likely to draw snappy synchs. In May and June, the flash of these lightning bugs, the call of barred and great horned owls, and the eyeshine of alligators are all possible experiences on this 1.75-mile boardwalk, which is wheelchair accessible. Preregistration is required.

Congaree National Park, less than twenty miles southeast of Columbia in the South Carolina Midlands, hosts a lottery like the one in the Smokies and issues a small number of spots for their firefly viewing sessions, held on eight days in May. The front-country trails are closed to hikers at 4 p.m. on firefly dates. Since Congaree is much warmer than the Appalachian sites, the lightning bugs show up sooner on the calendar. The lottery takes place in April, and passes are awarded in advance through the Congaree website. One designated firefly trail is the sole venue, and participants are restricted from using bug spray, phones, or white flashlights during the viewing sessions.

Congaree, which is the largest contiguous old-growth bottomland hardwood forest in North America, is an official UNESCO International Biosphere Reserve, meaning that it is part of a partnership among diverse organizations that work with local citizens to develop sustainable uses of this natural asset. More than 15,000 acres of Congaree National Park are federally designated wilderness and serve as the subject of collaborative research by experts from the UNESCO network.

In 2019, naturalists at another UNESCO Biosphere Reserve—Grandfather

Mountain in North Carolina—revealed the presence of synchronous fire-flies at elevations higher than previously known. Mikey Woody, a native of Lansing, North Carolina, who was working for the Grandfather Mountain Stewardship Foundation that summer, went out to watch the sunset at Blue Ridge Parkway overlook, on the land of Grandfather Mountain. At dark, she saw synchronous fireflies and reported it to the foundation and to parkway officials. Soon thereafter, Clyde Sorenson, a professor of entomology at NC State University, confirmed the finding, when he came to give a program on pollinators for the foundation. Sorenson stayed at the guest cabin on Grandfather, located at an altitude of 4,200 feet. He saw *Photinus carolinus* and the femme fatale fireflies, *Photuris versicolor*. Sorensen's initial findings led to more formal firefly surveys on the mountain. Grandfather reaches an altitude of nearly 6,000 feet, and the surveyors documented how firefly flashing slows at higher altitudes because of lower temperatures. At one overlook, the observers witnessed more than 1,000 fireflies on a single evening.

Yet another firefly species, called the blue ghost (*Phausis reticulata*), has become a hot ticket at several sites around Asheville in the North Carolina mountains. Guided viewing opportunities have been organized in the past few years in Pisgah National Forest and in Dupont State Recreational Forest. After word got out that these unusual beauties were visible from many trails at night in late spring and early summer, DuPont Forest had to restrict access on some pathways after dark during the mating season to guard against damage to the blue ghost populations.

Blue ghosts are distinguished by their longer period of illumination—up to a minute or more. Males shine with a faint bluish color and travel about one to two feet above ground. They tend to fly slowly in a meandering pattern, though sometimes they pick up the pace as they aim for their target —the *Phausis* females on the ground.

Females of this species become adults but never develop wings. Instead, they crawl in the leaf litter with a very limited range. To signal the males, they present with tiny glowing dots that run the length of their small bodies, which are less than four-tenths of an inch. The species flourishes in moist and mature leaf litter that has gone undisturbed for long periods. They tend to be present in numbers that make it easier to see them for only about two to four weeks every year. Timing of breeding depends on temperature, moisture, and elevation.

To experience the blue ghosts, we signed up for an evening hike led by Tal Galton, who runs Snakeroot Ecotours, out of Celo, North Carolina. Celo is home to a summer camp and a private residential school for grades seven through nine that emphasizes respect for nature and community. Celo rests in the pastoral Toe River Valley below the Black Mountains (North Carolina's highest range), which also happens to be the dominant view to the south at my 150-year-old log cabin, which sits at an altitude of 3,500 feet.

Neighbors recommended Tal as one of the best nature guides in the area. Year-round, he offers full-day and overnight outings, nature retreats, youth programs, and nature walks, along with seasonal hunts for orchids, mushrooms, fox fire, and fireflies in locations throughout the Pisgah National Forest and sometimes beyond.

A half hour before dark, we met our small tour group on the side of a private road and signed a form promising not to reveal the location, out of respect for the landowners and the fireflies. The week before had been wet, so our prospects were good, Tal said. Wearing a ball cap, a flannel shirt, and shorts, Tal is clearly a creature of the outdoors, easy with sharing his knowledge and enthusiasm for the region, where he has lived for twenty-five years.

It was late May. A few cows grazed at the far end of a bright-green field beside the gravel road where we began. Tal said if it were a bit later in the summer, we'd likely see some synchronized lightning bugs. Their mating season sometimes overlaps with the blue ghosts, and both are forest dwellers. He also mentioned the possibility of seeing a few other firefly species that are more at home in open meadows and fields when we returned to this spot later in the evening.

We followed Tal into the forest that flanked the gravel road. He told us that the area we were entering had been logged heavily ninety years before, but the 1,100 acres surrounding us had grown back, largely undisturbed, which is one reason the lightning bugs are so prevalent here. "Given enough time, species can recover, and in Appalachia, in general, we have much more forested land than was here a century ago," Tal said. He's seen hundreds of blue ghosts flying in an evening, but they don't stay out long—maybe an hour or so after darkness falls.

Dr. Jennifer Frick Rupert, the Dalton Professor of Biology and Environmental Sciences at North Carolina's Brevard College, began researching blue ghosts nearly two decades ago and is responsible for most of what we know about the species, Tal said. "When you realize how briefly they are visible to be studied—maybe forty hours a year during their mating display—you understand why our knowledge of some fireflies is so limited. What we *don't* know is much greater than what we do know. Of course, this applies to everything in nature."

*A Blue Ridge meadow at dusk will soon erupt with at least three bright species of fireflies—synchronous, flash-train, and triple flash. Deeper in nearby woods, blue ghost fireflies will also emerge for a short time when true dark arrives.*

As it grew darker under the tree canopy, Tal cautioned us to watch for puddles and uneven ground. He encouraged us to snuff our red flashlights and simply take our time and step carefully. This was a familiar practice by now. Our trek was not long. We stopped within earshot of a stream off to one side of the road, burbling under the last calls of a wood thrush, the songbird that, to me, sounds like no other, as if it had a throat made of two pan flutes playing in harmonic thirds.

We had brought nothing to sit on for this excursion, so we stood quietly as the dark swallowed us. The glimmer began on the opposite side of the road from the sound of water. Seeing lightning bugs that did not wink but kept their lights on for many seconds was a different sensation. They weaved low among the trees, rose, and dipped around the underbrush, ever determined to find their target, a dimly glowing female. It reminded me of a monster movie, an old, black-and-white film that had a scene in which the townspeople rallied to find and destroy the monster—maybe Frankenstein or the Wolfman. The mob took off through the woods, carrying burning torches and wandering back and forth among the trees, eager to confront the beast that had tortured their village.

When I saw the blue ghosts move, I was instantly reminded of those townspeople, lights weaving this way and that, hoping to uncover their nemesis. The blue ghosts, however, were not menacing. One woman in our group said they were just like flying fairies. Someone else blurted out "Tinkerbell!" The whole scene might have been aptly accompanied by music from Disney's *Fantasia*.

Tal explained that in Iceland, citizens to this day maintain a long and deep mythology about elves and fairies. He said the country has laws on the books that forbid developers from disturbing elf habitat, which is still marked on maps. According to a 2013 story in *The Guardian* newspaper, elf advocates in Iceland "joined forces with environmentalists to urge authorities to abandon a highway project that they claim will disturb elf habitat, including an elf church." The article goes on to illuminate how Icelandic people, especially in their isolation, remain especially close to the land. Development projects that threaten history and cultural treasures in addition to environmental integrity are often legally challenged in multiple ways.

"To me, elves are a metaphor for all those living things that are hidden from our view 99 percent of the time," Tal said. "Blue ghosts fit the elf metaphor perfectly. They are permanent residents of the landscape, mostly invisible to us, but during those fleeting moments—the one percent of their lifespan that they come to our attention—they are absolutely magical. Disturbing these woods would devastate the fireflies. The females can't get up and fly to new spots, and their range during their lifetime may be no bigger than the porch on my house."

Eventually, Tal took a small group toward the stream on the opposite side of the road to see if there was more going on in that direction. Somehow, I missed that they were looking for Dismalites, which Tal later told me are sometimes visible on the stream bank. But I could not get enough of the blue ghosts in that moment. In retrospect, I still had not realized that the promoted "rarity" of some bioluminescent creatures—perhaps even the Dismalites—is either a human invention or pure blindness to what is around us.

Because they stay lit for nearly a minute, images of blue ghosts have been successfully captured with digital cameras using a tripod and an open camera shutter, prolonged to capture more light, but the blue/green light is faint and still must be digitally enhanced. In pictures, the firefly is not as it appears to the naked eye. People also see colors differently, the firefly expert Lynn Faust tells us. Younger eyes see the color more accurately. I was simply glad to see them in their habitat. There is no substitute for that.

As Tal predicted, when we returned to the cow pasture, the field was lit up with fireflies at several levels—most strobing up toward the trees. Wild streaks of bold yellow burned across the middle of the field. These were flash-train fireflies that can zip and blink through the air almost as quickly as shooting stars. What Lynn Faust calls "flashbulb fireflies" exploded in the treetops. Both fireflies are classified as the species *Photuris versicolor*, but less is known about them because they fly so high and are hard to catch.

The word that came to mind was "paparazzi." The display felt like an assault of flashing lights. Looking to the treetops, I wasn't sure if we were seeing fireflies or satellites crossing the sky. It was a splendid finale. All of us stared for a while, breathed in the cool night air, and then began to say our goodnights. We thanked Tal. He seemed pleased with the evening.

The next summer, near the end of June, I went alone to experience one of the three special nights when the nonprofit Grandfather Mountain Stewardship Foundation offers a new nature program called Grandfather Glows. On the last curve up Highway 221 that winds from the historic resort town of Linville, North Carolina, to the gate at Grandfather, a string of cars was already backed up. Staff in yellow vests checked tickets and gave parking directions along the grassy meadow where a long-running celebration of Scottish traditions, the Highland Games, is staged for a few days each summer.

This was only the second year the foundation had offered this evening event for viewing fireflies. Demand had been so great the first year that the registration process crashed the foundation's website. Now the park had enhanced the ticketing system and wisely scheduled alternate rain dates for each of the three evenings. They released 600 passes for each night, and attendance was near capacity for all but the first session, which had to be rescheduled on the rain date. As Landis Taylor, the foundation's vice president for marketing and communications, told me, "Forecasting the peak of firefly season more than a month ahead of time is where things get tricky. Who would have guessed that we'd have such a chilly May and June?" The elevation at Grandfather also throws the firefly prediction into a different category from other sites in the region, where firefly viewing comes earlier in the summer.

Grandfather Mountain, originally called Tanawha by the Cherokee people, is a few miles off the section of the Blue Ridge Parkway that took the longest to complete (in 1987) because of the advanced engineering required to create a cantilevered road skirting Grandfather's radical slope. The current name of the mountain derives from the dramatic profile of a bearded man, presumably sleeping, which is visible from several angles in the region.

When the French botanist André Michaux first climbed the mountain, in 1784, he declared it the tallest peak in North America. Though his calculation was wrong, Michaux's reports of the exceptional biodiversity on the mountain led to ongoing visits over the years by many naturalists, including Asa Gray and John Muir.

By 1952 Grandfather had been acquired by a prominent North Carolina family who ran it for decades as a travel destination for paying guests. The attraction included a paved road to the summit, wildlife exhibitions, a nature museum, a dizzying swinging bridge near the summit, and various hiking trails and nature programs. In 1992, in conjunction with The Nature Conservancy, Grandfather was designated a UNESCO Biosphere Reserve, and the family agreed to protect this natural asset and its unique ecosystems. In 2008, the State of North Carolina bought 2,700 acres of Grandfather Mountain for a state park. Some 750 acres, which include the public visitor sites, are under the protection of the Grandfather Mountain Stewardship Foundation, a nonprofit that is charged with studying, preserving, and educating the public about the sixteen distinct ecological communities and the seventy-three rare or endangered species on the mountain. In all, The Nature Conservancy, which holds conservation easements on nearly 4,000 acres here, will protect the watershed in perpetuity and prohibit any future development that would change the character of Grandfather.

The meadow at the base of the mountain was large enough that the hundreds of people in attendance at Grandfather that night seemed manageable. Tents had been set up to provide information about the different species of fireflies that live on the mountain and how to make your homeplace more lightning-bug friendly. Grandfather Mountain merchandise was available to take home. The tent with the longest line was the one where guests who needed red cellophane to cover their flashlights and cell phones could get a free supply. Foundation staff were scattered about to answer questions. A storyteller set up in one area to tell tales of firefly lore to children. Some families brought picnic suppers. It would be more than an hour before a school bus was available to ferry those who didn't want to hike up to the viewing site.

John Caveny, the director of conservation and education at Grandfather, spoke to the gathering at the beginning of the evening and took time to talk to me a bit as things were unfolding. He told me that the synchronized species (*Photinus carolinus*) had been found and recorded at Grandfather as early as the 1950s and 1960s, but staff did not spend time on the mountain at night. Nowadays, the Stewardship Foundation property is closed in the evening with twenty-four-hour security, and only thirteen primitive, hike-in camping sites are in the state park on the opposite side of the mountain.

Once Clyde Sorenson, of NC State University, confirmed the presence of the synchronized species in 2019, nine more bioluminous species were quickly identified on Grandfather, including blue ghosts and *Orfelia fultoni*, the Dismalite glowworms. Sorenson thinks another ten species are still to be discovered on the mountain. Research continues.

John said his team saw what they believed were blue ghost females glowing from under rhododendron bushes along the paved road to the summit, but on further inspection, they realized they were seeing *Orfelia fultoni*, instead. "We then began looking for the *Orfelia* and found them along the park road in the mossy undercut banks. It didn't seem a likely place," John said, "but apparently these banks stay damp enough to support the glowworms." Big dipper fireflies also hang around the mountain for a long period each year. The presence of these beloved firefly species has led the foundation to be even more selective in pesticide use to control invasive species on the mountain. They have changed their management practices when it comes to leaf blowing, John said.

Caveny, whose father was a game warden, was born in rural Jackson County, North Carolina, where the *Orfelia fultoni* were discovered, in 1939. Later, he lived in the foothills town of Elkin, where there was very little light pollution and plenty of lightning bugs, which he enjoyed as a boy. He went on to study at Appalachian State University, in Boone, and recently completed an advanced degree in natural resources at Oregon State University, all the while working at Grandfather, where he has been on staff for a decade.

Aliyah Albadri, a student at Lees-McRae College and a part-time interpretive park guide, came to the microphone a bit later and introduced the species we'd be seeing on the mountain that night. She also gave us some information about various cultural interpretations of fireflies around the world. The Greek word for "firefly" literally means "the one with the glowing butt," she said. In Borneo, fireflies are believed to be the spirits of the dead. In the Andes, they are characterized as "the eyes of ghosts." The ghost connection, Aliyah told us, makes sense, in that many old cemeteries are places where larger numbers of fireflies gather, because there is less light pollution and more mature vegetation there.

As the sun moved lower in the sky, folks began lining up to climb the grassy path from the meadow to the viewing site. By the time we were moving, I was glad for the red lanterns mounted low on posts along the way. It would be tough coming down later in the dark without some illumination. We were headed for a nearly level grade of pavement on the road (now closed) to the summit of Grandfather. We would sit in folding chairs or on blankets that we had been instructed to bring with us that night.

I walked along the pavement and placed my folding chair where I'd be looking up a bank and into a forested section of moss-covered trees with a fairly low understory. Others chose the downhill view on the opposite side of the road, including a group of photographers who set up their tripods and were getting ready to cover their heads and cameras under black drapes. The center line of the road was kept clear so the spectators and staff interpreters could move about.

I sat next to a young mother, her friend, and a small boy about four. He was wearing shoes with heels that lit up with his every step. "Lightning bug shoes!" I said to him, and he nodded vigorously. I chatted with his mother and her friend and told them about seeing the glowworms in Alabama and how I was told we might see them right here on this bank.

We waited. The moon rose from behind the slope in front of us, but it was mostly obscured by the tree canopy. People took their time settling in, and the folks who arrived by bus were still coming through in groups, looking for open spots to settle. I could smell bug spray being applied—not good for the insects we wanted to see. I saw others who had brought the patch form of bug repellant, a better choice perhaps. There was a good breeze, so mosquitos weren't really a problem that night. The organizers had said to prepare for cold, and I could already feel the temperature dropping.

The glow began on the ground. A female lightning bug came on and off right in front of my feet in the leaf litter. As in Tennessee, the show was slow to build, but eventually we could make out some synchrony. The boy with the lightning bug shoes played with his red flashlight, which was distracting. Finally, his mother picked him up and carried him in her arms for a walk up the road, so his shoes would not flash. When they came back to their spot sometime later, she told me she had seen the glowworms. "Uphill where the road curves to the right," she said. "Incredible. I thought you'd want to see it." I thanked her and left my chair, my notes, and my bag and wove my way quickly through the strolling observers to find the Dismalites.

It was different from being in an Alabama cave of very wet rocks. The glow seemed more like the glint of mica, but no less magical. I was able to get very close to the jutting rocks with their points of light. The cut in the bank was right on the roadside, a spot that in daytime no one would suspect was so alive—an elf habitat! People could have easily reached out and touched the moss-veiled rocks, but they did not. Many could have missed it altogether, it was so subtle, given the light levels from the rising moon. On the mountain's website, Grandfather's naturalists have noted that the blue light emitted from *Orfelia fultoni* is "the purest blue color of any terrestrial animal."

I stayed for another half hour or so, my puff vest zipped against the steady breeze, watching the fireflies smear their bold lights across the night in front of us. It took an hour to drive back to the cabin. Suddenly lit by my headlights, a giant owl swooped across the Blue Ridge Parkway to pounce on something. I was not sleepy in the least but energized by the presence of so much life in the dark.

I don't know why it never occurred to me to look for blue ghost lightning bugs in my own woods at the cabin. A friend who raises sheep in the same county offhandedly mentioned at a dinner party that he had blue ghosts in his woods. So, a couple of nights after the experience at Grandfather, I stayed outside later than usual, standing beside a long bench, my arms resting on the edge of the hog-wire fence that divides the garden and the woods on my little postage stamp of steep land. These woods have been relatively undisturbed since 1968, when the old cabin was disassembled on a mountaintop, numbered log by log, and dragged down that mountain by mules. It was reassembled here by the first owners and sited beautifully in relation to sun and moon. Some trees have fallen in my thirty-three years here, and I have left them alone. The wind blows hard in winter. Several broken snags down the hill have historic tiers of pileated woodpecker holes. The leaf litter on the forest floor is deep.

I could hear some fireworks down the ridge, an annual custom of the ranchers in the hollow who stock up and shoot them for several nights in early July. I sat down on the bench. I realized then that I have waited for darkness for many nights on this journey to learn about lightning bugs—often enough that it now feels like a new practice and a good one. I have also attended to swarms of passing mayflies and random moths, the last birdsong of the day, the odd yellow leaf dropping from the big old cherry with her three trunks here in the yard. White pine needles have been flying on the breeze coming from behind me. Right now, yellow lightning bugs are lifting and dropping randomly through the woods, probably big dippers, now that it is July.

I knew it was too late in the season for blue ghosts, but then I saw one, ten feet away, faint and pale, rising from the duff. Then another and another. They stayed lit for long stretches. They wove a drunken path over the leaves, looking for females. I could hardly contain my delight. Why didn't I ever look before?

Because I didn't know before.

In that moment, I realized, I was graduating from lightning bug school. I called Donna and my brother to come outside and join me. We three sat on the long bench. The blue ghosts flew toward us and around us, coming through the wire fence. One landed under the bench beneath my brother and continued glowing.

Over the next two nights, we got sharp enough to spot even the larvae in the duff, lit by their flashing dots. The vision got even better after my next-door neighbors left town and no light shone through the woods behind or beside us. On the last night of my blue ghost watch, July 4, I went out alone. Their numbers were dwindling, but I saw a few blue ghosts. High fireworks were also sparking up from the valley below. As I rose from the bench to go inside, I looked to the sky and the Big Dipper was overhead. Fireflies, fireworks, and stars—what could be better?

I wrote Tal Galton and asked if I could come see him again. One early morning a few days later, he and I sat on his porch in Celo, and Tal explained that the blue ghosts do sometimes have what is called a second flight in a year. No one knows what it means or what it contributes to the species' survival. Tal then smiled and told me I was one among dozens of others—friends and strangers who had lived in these parts for thirty to fifty years, some for generations—who had taken his lightning bug tour and then called him a few days afterward to tell him with great enthusiasm that they had seen blue ghosts in their own yards. They had learned to look, and they saw what had been there all along.

Now I have a new ritual I will practice each year, starting in the last week in May—a new way to mark the season, a new awareness of the "elves" that occupy these woods that do not really belong to me, but which I will protect more fiercely. And I will joyfully share the chance to see them with anyone who comes to visit during blue ghost season.

# 10

# Purple Martins

*History shows that it is not only senseless and cruel,*
*but also difficult to state who is a foreigner.*
—Claudio Magris

The annual migration of purple martins between the eastern United States and South America can be so dense with birds that they often show up on weather radar. One recent account claimed that the Doppler footprint of the flocks coming through South Carolina was wider than Hurricane Hugo had been in 1989. The martins come to nest in the United States in early spring and begin returning south midsummer.

In this century, purple martins' travel patterns in the South are probably the closest thing we'll ever see to the spectacle of the eighteenth-century migrations of the now-extinct passenger pigeon. John James Audubon described the blanket of those bulky birds that he witnessed on a trip to Louisville, Kentucky, in 1813: "The air was literally filled with Pigeons; the light of noon-day was obscured as by an eclipse." The train of birds continued overhead for three days straight. Audubon estimated that he saw a billion birds. Because their plumage was prized and their clumsy presence damaged trees where they landed, passenger pigeons eventually suffered a fate of eradication by hunters at the turn of the twentieth century.

The lithe and handsome purple martin (*Progne subis*) is the largest member of the swallow family in North America. Its annual migration is still celebrated widely. Come July, in certain spots in the South, hundreds of thousands of martins gather to swoop, swirl, and rest at sunset. At sunrise, they take off once more in a cloud to fatten up for their eventual return for half the year to destinations south of the equator.

For better or worse, in the eastern United States, purple martins have become completely dependent upon humans for their nesting and breeding habits. When they arrive here en masse in early spring, they no longer set up housekeeping in tree snags, abandoned woodpecker holes, or other natural cavities. Purple martins, which tend to be more communal than most birds, have come to expect purpose-built housing, including fanciful apartment complexes and condos that human beings design and maintain for their guests. They allow people to touch and feed them. Being a "purple martin landlord," as these hobbyists call themselves, has been a passionate ritual for many rural southerners, but conservation groups are concerned that these dedicated human advocates are aging out without sufficient recruits to replace them.

In the open countryside, you've surely seen the tall poles. They're festooned with an assemblage of white, gourd-shaped containers that are drilled with an opening and hung at the ends of what could be the ribs of a round umbrella without a canopy. For centuries, these housing complexes have been a common sight in much of the South, at least below 4,500 feet in altitude. (Purple martins don't generally hang out at higher elevations.) And now, with internet-based interest groups monitoring colonies and maintaining region-wide communication, it's easier for followers to learn and share the latest whereabouts of the migrants.

Purple martins enjoy "larger bodies of water where reed beds and dry islands with low, thick brush provide sanctuary from predators," says the national Purple Martin Conservation Association, headquartered in Erie, Pennsylvania. Waterside destinations, the PMCA website explains, provide "a micro-climate warmer and less windy than land."

On the day of our reservation to see purple martins roosting for the night on an island in Lake Murray, we drive into South Carolina on Interstate 26 and exit at Irmo, heading toward Lexington, a town that sits on the west side of the Broad River. On the east side of the river is Columbia, the state capital. All the way from Irmo to Lexington, swaths are being cut through the piney landscape where high-end housing developments and shopping centers are taking over.

By contrast, the windows of cookie-cutter clapboard cottages from the 1940s, now closer to the widened roads than before, are shaded by old-fashioned lapped aluminum awnings installed before the prevalence of air conditioning. Brick ranchers from the 1960s are marked for demolition. The entrances to the upscale communities under construction are flanked by colorful banners beckoning house hunters with a yen to live near the lake. As is common in expanding population centers in the South, the proximity of rural poverty and new money is striking.

As we go up and over the massive span of the Lake Murray dam, the skyline of Columbia rises in the distance. When this earthen dam was built in 1928, it was the largest in the world. According to a visitors' guide, Lake Murray's shoreline has a circumference of 650 miles. The lake is among the state's largest, covering 50,000 acres. It's fourteen miles wide and stretches forty-one miles from one end to the other. In addition to well-heeled house hunters, it is irresistible to migrating purple martins.

Like so many impoundments created in the early twentieth century to produce electricity, preparation for the lake required massive logging and tree removal. Some 2,000 workers performed the labor, using only cross-cut saws and axes. They produced the raw material for 100 million board feet of lumber. Some of the lumber was used in the dam itself and some to construct three miles of the railroad line that runs alongside it.

A dozen communities were abandoned when the water began to rise on August 31, 1929. Residents received between fifteen and forty-five dollars an acre for the property they were forced to leave, a local website says. Six schools and three churches were relocated. Nearly 200 graveyards stood in the path of the rising reservoir. More than 2,000 caskets were moved, but not all. The guide does not mention the wildlife or botanical ecosystems that once occupied the forest here. In 2005, a backup dam—wiping out still more creatures and habitat—was built for $275 million. The federal government required the additional dam to prevent a devastating flood of the capital city and points south in the event of an earthquake, but some still question whether the second dam is adequate to prevent such a disaster.

Cotton boll clouds are scattered across the South Carolina sky; it's a perfect day for a boat ride. Enough breeze comes off the lake to flap the flags on the parked pontoon boats and jet skis at the marina when we arrive. The vessels pull and clang against their moorings in the maze of docks at Jakes Landing. Though summer high season is well under way, the boat storage facility is mostly quiet on a late Wednesday afternoon. I step inside the marina store to confirm our arrival for a two-hour sunset tour. The usual gear—life preservers, boat hooks, bait, tackle, drinks, ice, oil, and batteries—are haphazardly arranged among snacks and stacks of T-shirts. A young woman at the cash register sells me some gum and bottled water. I go back to the car to wait for our boat captain to arrive.

More than mere songbirds, purple martins are aerial acrobats. They can snag an insect meal in flight or swoop low over water for a quick drink without landing. Their beaks open wide. They can even bathe on the fly. Their broad-chested, streamlined bodies are flanked by tapered wings that allow them to glide for restful stretches. In flight, their tail feathers spread out like a Japanese fan. The males' glossy indigo feathers are iridescent. The brownish females and juveniles have a white and brown speckled chest and strokes of inky blue and gray on top of their heads.

Purple martins are passerines, meaning that their feet allow them to perch with excellent posture. Nonpasserine birds have feet adapted for other purposes, such as paddling or catching prey. Examples of the nonpasserine group are ducks, swans, woodpeckers, hawks, owls, and eagles. Hummingbirds are nonpasserines, too: their weak feet and legs don't allow them to walk around on the ground like other birds. Birds of prey and woodpeckers certainly can hang on a branch with their talons, but they are not adapted for much walking.

Our adventure on Lake Murray will answer the question, *What do half a million purple martins look like in flight?* It will also raise questions about the costs to birds when humans get involved in their life patterns.

A man comes out of the marina store, pushing the door with his backside. He wears a safety yellow, short-sleeved shirt and matching visor. He balances a clipboard and cellphone against his chest with one arm. His other arm, I notice, ends right below his sleeve. I approach him. Deeply tanned and quick to smile, Rick Crout puts down his gear and pulls out a business card for me. He is the proprietor of Captain Hook's Lake Murray Tours. An aviator for forty years and now retired, Rick explains that he's the go-to boat driver for bachelor and bachelorette parties, birthday celebrations, and pleasure cruises on the lake. He also collaborates with Zach Steinhauser, a certified Coast Guard captain, filmmaker, and wildlife tour guide, who organizes the purple martin excursions here at Jakes Landing.

Captain Zach hosts many other nature tours year-round, the most unusual being snorkeling trips in the chilly French Broad River to see endangered hellbender salamanders. He and Rick divide up the schedule of guided trips here at sunset and sunrise to Bomb Island, where the purple martins roost in July and August. They open reservations on June 1, and the tours sell out quickly. Depending on weather, they'll conduct thirty-six afternoon tours from the end of June to the end of August and a few sunrise tours on weekends through the season. Other vendors in the area also provide tours, and it's possible to bring your own boat or rent one to get out and see the birds.

Rick says that getting to Bomb Island, near the center of the lake, will take us about a half hour if we go full speed, but he promises to extend the excursion and show us a bit more of the shoreline and other islands, giving the sun time to drop farther toward the water. Our group of six soon assembles. None of us have seen the martins here before.

Bomb Island is so named because, in the early 1940s, bomber pilots from a nearby army air base trained in B-25 airplanes over the lake, releasing practice ordnance on seven of Lake Murray's islands in preparation for World War II missions. Before it was over, five planes ended up in the lake: some crashed; others were ditched by the pilots. Four were salvaged during the war and the last one was retrieved in 2005.

Rick leads us to the dock, and after signing release forms, we introduce ourselves. We have all brought bags that hold cameras, binoculars, and snacks. One couple, maybe in their sixties, admits to having just begun dating recently. They have driven across the state from Charleston and are acting like teenagers. They take the bench seat in the back of the boat and cuddle. The other twosome are veteran birdwatchers, recently back from Machu Pichu, they say. The stern man has a heavy telephoto lens. His smaller partner dutifully takes a seat that will keep her out of the line of sight for his shots.

Donna and I settle in the bow. This luxurious pontoon boat, which Rick casually mentions is worth six figures, has impossibly perfect beige seats that would seem to be leather. However, as I found out later, they are "ultra-soft marine vinyl." The boat also features abundant "Yeti-ready" cup holders, USB ports on every bench, and a sound system that easily defeats the roar from the dual inboard motors. The brand name, Barletta, seems fit for a driver who has named himself Captain Hook.

Rick unfastens the boat and takes the wheel and we're off, slowly at first, through the no-wake zone. He favors contemporary country music and only lowers the volume for a little while to give his spiel about purple martins, their habits, and what we'll be seeing.

The dam is behind us, massive and towering. On the other side of the boat, we are closer to the shoreline and a row of expansive houses—many brand-new—built high above the water, with graceful landscaping running down to private docks where jet skis and powerboats are waiting. Some of these houses also have purple martin houses in the yard—gourd-shaped but made of molded white plastic.

*Purple martins have been celebrated for centuries across the rural South as they migrate in and out of the region in summer.*

In 1831, John James Audubon reported seeing martin boxes at "almost every country tavern" in his travels, and ever the bird-backer, he suggested that the better the box, the better the business at the tavern. Nowadays, some towns across the eastern and central United States promote their purple martins with summer celebrations, often led by local Audubon chapters. Others have adopted purple martins as part of their municipal identity.

In Rutherfordton, North Carolina, the purple martin was part of the family crest of the town's namesake, the Revolutionary War general Griffith Rutherford. In 2015, the town established a Purple Martin Greenway, where a walking and birdwatching trail connects two public parks.

The South Carolina town of St. Matthews—the seat of Calhoun County and birthplace of actor Viola Davis—adopted the purple martin as its mascot and sponsors a festival each April with a parade, golf tournament, street dance, car show, and education sessions on native plants and birds.

Purple martin popularity has fostered single-state organizations that educate and fundraise for the birds' protection. These groups sell a prodigious range of nest box designs that sometimes mimic the architectural vernacular of a given area—from steep-roofed Victorians to Williamsburg colonials to country cottages. Today, in addition to house plans and building kits, martin mavens offer tips on proper housing placement in the yard and how to perform nest checks and protect the birds from predators and parasites.

The practice of building elaborate martin houses went commercial in the late 1800s, when the amateur ornithologist J. Warren Jacobs of Waynesburg, Pennsylvania, declared the purple martin to be his favorite among songbirds. After displaying his birdhouses at the 1893 Chicago World's Fair, Jacobs began mass-producing the structures, including a 105-room multistory "Capitol" birdhouse, six feet tall and weighing more than 500 pounds. Jacobs was also a poet, essayist, and collector of more than 10,000 bird eggs for his Jacobs Museum of Applied Oology, which often received international visitors. His bird houses sold around the world. Fifteen went to Henry Ford and six to William Rockefeller and at least one to Thomas Edison. The business lasted fifty years. His designs are once more being built in Waynesburg by three partners, who revived the company in 2001.

Purple martin landlords today rig their multiple-unit housing with ropes and winches to raise and lower the pole-mounted complexes for easy cleaning and inspection of nests and to count each clutch of eggs (usually four or five per nest). Baffles on the poles protect against snake and raccoon invasions. Some people install electrified predator guards.

In 2019, a team of biologists and social anthropologists studied this unusual human/avian relationship and reported that some landlords believed that the birds seemed to know their caretakers: "I'd hate to sound probably corny," one landlord admitted, "but it's a friendship. I mean, they're backyard pets, without being domesticated, really." "Synanthropy" is the scientific term for purple martins' preference for living close to humans.

Humans have taught purple martins to depend on them, and the birds, in turn, practice what biologists call "site fidelity." Like reliable members of the family, the same martins and their progeny will come back year after year to their individual breeding sites. The landlords wait with anticipation and delight in their arrival.

I've read stories of landlords who stockpile frozen crickets to serve arriving martins when early spring weather takes a bad turn. Heavy rain, sleet, and rapid temperature drops can keep martins from finding or catching flying insects. Martin lovers have been known to flip crickets in the vicinity of birds huddling against the weather to keep them from starving before warm weather arrives in earnest.

In the 2019 study, the social anthropologists interviewed twenty-four martin landlords—mostly older white men in the South. Only two were women. "The long-term bond between the birds and their human keepers meant that the landlords were confronting not only their own mortality but also the extinction of the birds to which they have devoted themselves," the authors of the study concluded. "Their continuing struggles to recruit younger generations into landlording suggest the need for new models of martin conservation that can appeal to the post domestic, and internet savvy sensibilities of today's youth."

Contemporary wildlife management generally begins with humans trying to protect or improve habitat for wildlife, but the dependence of purple martins on humans for housing has evolved over centuries. Historians tell us that the first martin houses were gourds grown by Indigenous people, who hollowed them out, removed seeds, and carved an entrance hole to attract the birds. According to a nineteenth-century ornithology book, Choctaw and Chickasaw people were documented putting out gourds for the birds as early as 1808. The Indigenous people apparently welcomed martins to reduce insect populations in their settlements and to discourage other birds, such as crows, from feeding in freshly planted fields. In the same era, African Americans living on the banks of the Mississippi were observed hanging gourds from long cane poles.

I'd always heard that purple martins eat great quantities of pesky mosquitoes, but studies dispel the myth. Martins are fonder of and better fed by larger insects, such as dragonflies, grasshoppers, beetles, flies, bees, and wasps. According to a recent investigation, purple martins also vacuum up fire ants and queens from their mounds in great quantities. Mosquitoes, being so small, can compose only a tiny percentage of the martin diet.

Even with such devoted caretakers, the North American breeding population of purple martins has declined by 25 percent since 1966, according to *Audubon Magazine*. The North Carolina Purple Martin Society has begun pressing its longtime landlords on its website to teach their younger friends and family members how to care for the birds to ensure a continuing legacy of colony maintenance: "There has been less of a trend of putting up housing for these birds in recent years and their populations have suffered as a result of lack of managed housing and nest competition with two non-native, aggressive, invasive species." Those invasive species are starlings and house sparrows.

As we head down the lake, Rick points out Captain Zach's house, where a similar pontoon boat is docked. He tells us that national bass tournaments are popular here on Lake Murray. Fishing is generally good—largemouth, stripers, bream, bluegill, and catfish. The man at the back of the boat asks if there are alligators in the lake.

Rick laughs. "No. Saw one on the golf course last week near Myrtle Beach. You probably read that story about the guy who was attacked." The man nods and pulls his date closer to him. She smiles. Her makeup is flawless.

We motor past a condo development with an imported beach of impossibly white sand. Most of the erosion spots on the banks are red clay. A mansion on a prominent hill in the distance looks as if it had been designed by Thomas Jefferson.

Rick eventually rounds a promontory of land and turns the boat away from the main body of the lake and into a peaceful cove. "Sometimes we see deer in here," he says, and throttles back to coast, the sound system still blaring. The man in the back says he might like to rent his own boat next time and take his lady parking in this pretty spot. TMI, buddy.

Grass is growing in the shallows, and as we drift toward the end of the cove, we see a bright green algae bloom. Water too warm. Fertilizer runoff, I am guessing, from the elegant lawns we just passed.

In addition to the decline in landlord numbers and the reductions in insect populations by habitat disruption and development in the South, the two invasive species of birds—European starlings (*Sturnus vulgaris*) and English house sparrows (*Passer domesticus*)—have become a vexation for purple martin landlords. These birds were introduced to North America in the 1800s and have been unpopular ever since. Neither are protected species in this country and "may be legally controlled by trapping or shooting, depending on your city ordinances," the PMCA notes.

Besides the occasional hawk or owl predator, starlings and sparrows have thus become the focus of multiple strategies to protect purple martins, leading to a raft of commercial products for hobbyists and other groups who have problems with these birds. And here, human intervention takes a troubling turn, at least for the softhearted.

The English house sparrow was imported to this country "to control a plague of inchworms on the eastern seaboard . . . While the sparrows initially appeared effective . . . the inchworms soon rebounded even as the sparrows exploded in population, annoying city dwellers with their droppings, messy nests, and incessant tuneless cheeping," according to the journal *Environmental Humanities*.

Like the purple martin, the house sparrow has come to prefer human architecture for nesting, including eaves, streetlights, bridges, and birdboxes.

They are aggressive and will kill purple martins and bluebirds to take over their houses. They will also push aside native birds at feeders. Note that these latter aggressions are created by people putting up bird houses and feeders, a charge to which I must also plead guilty.

The methods people use to fight back predators may include netting, sonic devices, scented repellants, and tactile gels—all adopted by purple martin lovers but also by businesses that try to discourage nesting house sparrows. Those birds you see inside large airport terminals? Most are house sparrows that have found a way to enter. It has always struck me as funny that the sparrows practice their take offs and landings while carrying bits of popcorn and French fries—the scraps from restless human passengers also waiting to take off.

Growing up, I never heard a kind word about European starlings from the bird lovers in my family. The common or European starling travels fast (nearly fifty miles per hour) and in tight flocks. They will overrun backyard feeders—the main reason for my grandparents' contempt. They will also invade other birds' nests and kill hatchlings. Like aggressive blue jays and mockingbirds, starlings can vocally mimic other species—thrush, hawk, robin, flicker, pewee—and they can quickly descend like a black cloud on a lawn or field, stabbing their bills into the ground in search of food. But starlings usually move on.

I have seen bird lovers who adopt a single, favorite species and then develop their own aggressive behaviors to defend it. Many years ago, I witnessed one of North Carolina's most avid bluebird advocates as he sat on his front porch shooting robins with a pellet gun as they came to feed on the live meal worms he'd raised in his barn to feed the hundreds of bluebirds he attracted and housed on his farm. Dead robins surrounded his platform feeder. He made sure his worms were only for the bluebirds.

The social anthropologists noted in their landlord studies that many younger folks do not have the inclination to kill other birds to protect the martins. I get it.

Soon, we turn back out of the cove, and Rick guns the inboard motors. He promises we'll begin seeing some birds. He is cruising straight ahead for a small island with a few dead trees near the shore. As we get closer, we can make out what may be a nest at the top of a weathered pine. "Ospreys," Rick says. We slow down and come close to see them. One of the three birds flies off as if to distract us from the nest. The cameras come out and the sightlines are good. Shutters are clicking. The osprey that left the nest eventually returns with a fish.

As we drift, a purple metallic party balloon lands in the water and is propelled by wind toward one end of the osprey island. It bobs away as the wake from a passing jet ski hits it. Now, in the distance, we can also see a few purple martins feeding, diving, coasting, and skimming the water out near the next island ahead.

Rick points. "That's Bomb Island. It's the only island where you can't land a boat. It's posted for roosting martins." As we motor closer, I can read the sign put up by Dominion Energy. The banks are steep and a raw red, eroding badly. The edges of the banks look like the landscape of Utah's Arches National Monument in miniature. Higher ground is thick with shrubs and trees. "The power company owns the island," Rick says. "We wonder whether they will protect it."

Starlings often travel in the company of red-winged blackbirds (*Agelaius phoeniceus*), common grackles (*Quiscalus quiscula*), and brown-headed cowbirds (*Moiothrus ater*)—other birds that my songbird-loving grandparents used to complain about. These birds tend to get the same treatment as starlings, especially for their offenses in cities. Noisy gangs roost in large numbers on urban buildings and leave their droppings behind when they move on.

Nature writers have perpetuated the tale that starlings in the United States are inbred, descending exclusively from a hundred birds released in Central Park in the 1870s. The famous naturalist Edwin Way Teale popularized the story that starlings were brought here as part of a peculiar effort by one man to introduce to the United States every bird mentioned in Shakespeare's plays. This myth now seems to be invoked every time a story about starlings is written. However, current scholarship argues that undocumented releases of starlings and escapees from the transatlantic pet trade were occurring much earlier in the 1800s in New York City and elsewhere in the country. In 1864, the commissioners of Central Park introduced several nonnative bird species, including the English house sparrow and the blackbird, both of which immediately thrived on the continent. The Cornell Lab of Ornithology figures that 200 million starlings now range from Alaska to Mexico. In addition to the nuisance they pose to buildings, they are the sworn enemy of corn and soybean farmers and ranchers who fatten their cattle in concentrated feedlots—a controversial practice. To counter damage by the birds, the animal feed company Ralston Purina developed a poison called Starlicide, which causes uremic poisoning and organ congestion in starlings.

Unfortunately, the poison can also harm bobwhite quail, pheasants, and rooks that take meals treated with Starlicide—another unintended consequence of human intervention. When the US Department of Agriculture deployed the toxin in New Jersey in 2009, the journalist Brian T. Murray described the scene in the *Newark Star-Ledger*: "Hundreds of birds that dropped dead on Somerset County cars, porches and snow-covered lawns, alarming residents over the weekend, were all of a rather foul breed of fowl —the notorious European starling, which the United States Department of Agriculture killed on purpose.... 'It was raining dead birds,' said Franklin Township mayor Brian Levine, explaining how people watched starlings drop throughout the Griggstown section of his town."

Nearly 5,000 birds died over that weekend, leaving residents to wonder if avian flu or West Nile disease had broken out, since no warning had been issued about the plan to rid the area of starlings.

Recently, in Nashville, Tennessee, even purple martins crossed over to the status of nuisance bird. On a string of summer evenings in 2020, migrating purple martins began roosting on and around the Schermerhorn Symphony Center. To the delight of some and the horror of others, the numbers of visiting birds increased the next year. Their droppings and the damage to trees around the once-pristine limestone concert hall eventually cost the symphony $100,000 for power-washing the exterior and sidewalks. The cleaning bill came due during the pandemic, when the organization had already furloughed its musicians amid heavy losses in concert ticket sales.

As the Nashville-based *New York Times* columnist Margaret Renkl explained to her readers, the assumption at first was that the offending birds were starlings, and a pest control firm was called in. Then conservation groups raised the alarm, explaining that the migrants were purple martins, a protected species. Volunteers kicked off a fundraising campaign to help the symphony with expenses.

The Schermerhorn then devised another offensive before the 2022 migration. They ordered that forty-one of their twenty-year-old lacebark elm trees around the building be cut down. The elms were approximately twenty-five feet tall and flanked the public entrance to the hall. Another public outcry, this time by tree enthusiasts, followed, and the organization then pledged to safeguard the ten remaining elms on the property and work with the Nashville Tree Conservation Corps to plant another 150 new trees around the city. Nashville applied for and became a designated Urban Bird Treaty City, based on a comprehensive plan of protection for all birds in the city. In 2023, bird-lover volunteers took to the streets to help explain to visitors and residents that the purple martins were back again in great numbers in a new spot near the Schermerhorn and were being carefully monitored and protected.

These stories designating certain species as the enemies of people and other birds show how we tend to assign human characteristics—heroic, villainous, invasive, greedy—to other species. Such vilification in our ecological narratives parallels the stereotyping and ostracizing through history that certain human groups have suffered—notably, migrants, according to Peter Coates, the author of *American Perceptions of Immigrant and Invasive Species: Strangers on the Land*. Human prejudices toward outsider species often cloud rather than clarify an analysis of what constitutes good public policy. All told, community efforts at human and wildlife management have tended to amplify the challenges.

As we come closer to Bomb Island, the martins, half a dozen or so at a time, come toward us as if out of nowhere. I can see them only when they are closing in, maybe the distance of a football field away. They are like a sprinkling of winged peppercorns in the air. Blackbirds are flying in, too —noisy, clumsier, a bit lower, and a few at a time, making a dash for the shoreline, where they land. They waddle around and confer. They know a good feeding scene, or perhaps they find safety in large numbers by migrating with the martins. The lake laps at their feet.

Now the frequency of waves is picking up as other boats slide in on either side of us, all keeping a respectful distance from one another and the island. The martins, however, are cruising the treetops when they get to the island. A pontoon vessel near us drops anchor, with that big thunking sound and a tall splash. Within minutes, three joyful kids jump into the water. The family's golden retriever is oblivious of the trickle of birds overhead. She paces the deck, wanting to join the children in the water as they splash about with Styrofoam noodles. An adult on the boat leashes the animal. Several boats around us also have dogs on board.

I finally ask Rick to turn off the music so that Donna can shoot some video and capture the sound of the birds, which is increasing. He is kind and quick to accommodate. Our stern photographer with the long lens reports an increase in numbers of birds in the distance. He can see farther than I can with my naked eyes or even my binoculars. For her part, Donna is shooting straight up. There are airplane contrails above, bright white chalk lines dissolving. The martins are high and pass in contrast to these trails, crossing the lines in the sky as if they are the early finishers in a marathon with a pace that's building. Trying to follow one bird and then another, I realize that the martins are coasting effortlessly and then winging a few beats and coasting again. It looks like fun.

Since 1989, purple martins have been coming to Bomb Island in late June to roost. Then in 2014, they didn't show up. Were they late? Delayed by something? Did the Fourth of July fireworks scare them away? Were they off course from South America? Locals worried. They kept checking for them. No birds.

It happened to be the year that the enthusiastic and geeky Adam Cole and a crew from *Skunk Bear*—National Public Radio's YouTube science show—came to do a story about the annual congress of martins. Cole, an infectiously cheerful storyteller who was not going to give up just because the birds were MIA, found and interviewed local martin landlords about their love for the birds. Then he contacted Julie Hovis, an endangered species biologist with Shaw Air Force Base near Sumter, South Carolina. Cole, Hovis, and one of the local martin lovers set out to find where the martins went. Hovis had helped tag some in previous years to track their winter migration habits.

Her first idea was to check local radar at sunrise. With Cole looking over her shoulder, Hovis managed to pick up what could be a cloud of birds taking off from an unnamed island on Lake Monticello, twenty-five miles north of Lake Murray, in Fairfield County. She and Cole took a boat out on Monticello that afternoon, and by sunset they had hunted the martins down, giving Cole a satisfying story. More publicity followed, which was good for business, said the local chamber of commerce.

Some speculated that the birds were spooked by a great horned owl that had been flushed out by agents of the South Carolina Department of Natural Resources on Bomb Island when they went looking there for recent changes that might have sent the birds elsewhere. Incidentally, they found evidence of a recent group of campers who had trespassed earlier in the season. The inspectors also discovered that Bradford pear and Chinese cherry trees had unfortunately taken root on the island. Both species of these pears and cherries are nonnative botanical invasives, and the seeds must have come in the droppings of the blackbirds that accompany the martins to their roosting habitat on Bomb Island.

Ultimately, the purple martins came back to Lake Murray the next year and have been alighting for the night every summer since. Julie Hovis was surprised, as was staff from the state natural resources department.

Disappearing roosting groups aren't all that rare, it turns out. Another purple martin colony that arrived for many years to occupy the underside of the William B. Umstead Bridge that crosses from Mann's Harbor, on the North Carolina mainland, to Roanoke Island, on the Outer Banks, did not return to its site in 2022. Interest in the birds had turned into a full-blown tourism activity on Croatan Sound, and the county built a pier at the west end of the bridge for better viewing of the roost. Beginning in 2007, North Carolina Department of Transportation officials also adopted a policy to accommodate the birds by dramatically lowering speed limits from fifty-five to twenty miles per hour on the bridge for the weeks the martins would be in residence.

When the birds didn't show up in great numbers, the DOT, ever hopeful, still lowered the limits. Sunset cruises to witness the martins were canceled, though. For some on the island, the disappearance of the martins echoed the long-unsolved mystery of *The Lost Colony*, an outdoor drama presented annually for more than eight decades on Roanoke Island. The play tells the story of the first English colonists who established a fort on the island in 1587 and then disappeared with scarcely a trace after three years. Their fate has yet to be fully explained. Just as mysteriously, the lost colony of martins disappeared for a season, but then returned to Mann's Harbor the next year to the delight of local naturalists.

The birds' numbers and noise are picking up, and more boats are arriving, creating a floating necklace around Bomb Island. The blackbirds are cackling. The martin calls are liquid, much more musical. The martins glide high above while the larger blackbirds flap along in phalanxes, some at eye level, all heading in only one direction: here. Our excitement is building with the changing tint of the sky.

This is a peculiar feeling, being on a gently rocking boat, looking straight up at a river of birds, the earth turning away from the sun. Rows of birds, now like a marching band on a football field, are moving in unison. They keep coming. They keep coming. Our companions on the boat begin uttering impulsive, startled "ahs." We laugh and shift our positions, craning our necks this way and that. We can't look at them all at once. The span of birds is too big.

"I've seen it night after night for years, and I never get over it," Rick says. He has begun backing up the boat toward the sun behind us. He takes us outside the circle of anchored boats and around the bend. The flocks are still thickening and moving in chaotic clouds, wave upon wave, above the island. Rick lines us up with the sun far out in the distance, sinking into the water. Now, the birds are dropping into the trees beside us, as if to signal time for bed, only to rise again suddenly, swerving and swirling in battalions. Then they are overhead, thousands deep. They seem to drift and drop, just above us but never too close. It's as if they know we are here for the show, and they are giving it to us. Again, I try following one bird in the swirl. There are some collisions and recoveries, quick dodges, turns, drops, longer coasting, swerving. The wind is blowing, too.

"Look at that."

"Oh, my god."

"That's crazy."

We are all laughing at them and at ourselves. It is, as one observer put it, a bird tornado. Now the birds are weaving strands of an open, crisscross fabric. There are as many of their dark bodies as there are openings of light that shine through the weave. The sky behind them has gone crimson with blue stripes, the birds overlaid like a loose shawl, lifting and dropping, lifting and dropping, as if a bed were about to be made. Finally, they tighten ranks—a panorama of black marks, just above the trees, a Jackson Pollock of birds. They are settling in. Their ranks are thinning. The sun is a poached egg, draining orange on the plate of the lake. We head back to the marina.

The next morning, we visit Wingard's Market, the local garden center that took our reservations for the boat tour. We hope to meet Zach Steinhauser, who has been developing a high profile across South Carolina as the chief advocate for the purple martins on Lake Murray. He fell in love with the birds as a child on visits to see his maternal grandparents who, more than fifty years ago, established a local nursery known for its selection of azaleas, now called Wingard's. When Zach's parents moved to Lake Murray, he was thirteen and already deeply interested in wildlife and photography. Zach became a volunteer with the nearby Carolina Wildlife Center and went on to earn a degree in wildlife ecology and conservation from the University of Florida, where he was active in red wolf conservation. In addition to his nature tours, Zach serves as Wingard's resident expert on native plants and wild birds. The sprawling operation is close to the south end of Lake Murray.

We spend some time wandering through the excellent collection of native perennials and notice a separate building for bird paraphernalia. First, we pick out a martin house made from a true gourd as a souvenir to buy, instead of one of the fancy, plastic houses. Then we study the plants some more. When we ask a young woman nearby about the zone hardiness of a particular peony, she gets on a walkie talkie and summons Zach.

Zach arrives wearing a lime-green T-shirt with the market logo, cargo shorts, and a brown ball cap turned backward. He is tan, blue-eyed, tall, bearded, and has just turned thirty. He has the ease and attentiveness of a person schooled in customer service from an early age. He tells us about the peony and then remembers my email message to him.

After four years of travel and filming, Zach has completed *Purple Haze: A Conservation Documentary*. The film carried him to Brazil, the southwestern United States, and several other southern states as he explored the contemporary challenges that purple martins face. Conservation organizations nationwide have been screening it. We chat about the birds and compliment Zach on our experience with Captain Rick.

I am still contemplating what the future might hold for purple martins. While they have won the affection of their landlords in this country, they have never had caretakers in their South American habitats, for better or worse. The six months they spend below the equator—from Colombia and Venezuela south to Brazil and northern Argentina—involves molting (getting new feathers ready for another migration) and feeding in tropical rainforests and on agricultural land. The Purple Martin Conservation Association and several universities have collected data on their journeys, tracking them using geolocation devices, conducting physical observation, collecting genetic material, and testing sample populations for pathogens and toxins.

Some martins have been found roosting in great numbers on the pipes at an oil refinery near Manaus, Brazil. They also huddle in urban areas on high wires and leave unwelcome droppings in cities. Scientists worry that deforestation and new river impoundments in the Amazon basin may alter martin habitat and that exposure to mercury mining and other toxins may affect the birds at their southernmost destinations. The ultimate impact of industrial pollution and climate change on the species going forward is a puzzle. Bridget Stutchbury, a biologist at York University in Toronto, has reported in a migration study: "Martins that feed over agricultural fields in southern Brazil are undoubtedly exposed to dangerous pesticides that are used in large quantities throughout most of Latin America." Some scientists have suggested that the forested regions of northern Brazil will offer the highest survival rates, but climate change will affect these areas even as deforestation continues. Northeastern South America is expected to get hotter and drier, and southern Brazil is expected to become wetter. Somehow, the birds survive in South America without the comfortable housing that landlords in the United States provide. Perhaps the birds will adapt to changes along the migration trip in the United States as landlords age out and martin houses fall into ruin. Zach wants to encourage new landlords to keep providing housing, but will that be sufficient? Would the birds be better off if left to their own devices?

The story of the purple martin is an extreme example of intentional human interference in the nesting cycle for a single species. Zach is taking his concerns and affection for these birds to the public through his documentary film. I asked him about killing starlings as part of caring for purple martins, and he agreed with the practice. But what about the starlings and the house sparrows?

The social anthropologists who surveyed the aging purple martin landlords in the United States suggested that new, younger landlords might be encouraged to use pest control businesses to service and clean the bird houses and to manage the predator starlings and house sparrows. That seems rather bougie and unlikely to me. The researchers also proposed "the option of moving martin houses and colonies into the public sphere, rather than maintaining them as the conservation activity of private homeowners."

As we have witnessed the magnificence of ordinary natural processes on this journey across the South, we have seen time and again how humans have altered so much in so many places, creating a mess. Meanwhile the martins persevere on their flight path, settling into tight quarters on a tiny island for another night, as we, too, settle in on this small-and-getting-smaller ball of earth.

# 11

## Moths and Butterflies

*The moth settled onto the curtain and sat still. . . .*
*No human eye had looked at this moth before;*
*no one would see its friends. So much detail goes*
*unnoticed in the world.* —Barbara Kingsolver

A shooting star cut a bright streak through the night sky. It was August, four o'clock in the morning, on a remote, two-lane road in rural South Carolina. Before sunset the day before, the temperatures had reached into the breath-defying high nineties. The temperature gap between inside and outside the car was still a chasm. That might explain why the windshield kept clouding up. Swirling patches of fog rushed against the car like ghosts. I kept switching the wipers on and off and the headlights from bright to dim to see the way ahead. I sent a silent prayer of thanks to the inventor of those orange reflector dots that defined the median. And I gave thanks to the road builders for installing them out here where no towns or gas stations showed up along the route for many, many miles. Only a couple of churches offered artificial light at this hour. It was Sunday, after all, and we were crossing the state on impulse from west to east, aiming to arrive in time for sunrise at the coast, hoping to find a few roseate spoonbills feeding in the marsh before the day got too hot.

It came to me that night what a long time it had been since I had driven a dark road in the rural South—partly because fewer and fewer dark roads without development exist, and partly because I don't go out so late anymore. I needed to be reminded how much insect life is still active in the deepest night in this region's rich forests and fields. I had forgotten the sheer volume of flying things that could be drawn to passing car beams. Moths showed up in my headlights and drafted over and around the windshield, mostly without impact. Some looked like flying guitar picks, some smaller, like confetti. Once, something invisible hit the windshield hard, without leaving a trace. Perhaps a bat in pursuit of the moths?

That night I thought of Joseph Merrill Lynch. He's a North Carolina native who's had a long career with The Nature Conservancy. He's now a consulting biologist on projects with the National Park Service, the US Forest Service, the Conservation Fund, and other nonprofit groups. A few months before this jaunt to South Carolina, Merrill and I had met in the mountains to talk about moths.

"Moths will come to you," he said to me with delight. "You don't have to pursue them, and if you begin to study them, you will be rewarded." He took a sip of coffee. "Every month of the year, something is flying, even in January." Merrill and his family live close to the state line between North Carolina and Tennessee. In the time they've lived there, Merrill says he and his now teenage daughter Dovie have photographed and identified 1,300 species of moths in their yard, along with a few still unidentified. Merrill has ratcheted up his macro photography skills because of moths.

"Every species of moth has a host plant or multiple host plants," he explained. Some moths are picky eaters, depending on only one plant for food in the larval stage, and other species graze on many hosts as they move toward adulthood. "These plants include ferns, grasses, evergreens, and hardwoods," he said. "Some caterpillars eat flowers, some prefer leaves —often young leaves—and some eat the fruit. Some larvae live in the roots of plants and emerge to become beautiful flying moths. And some species are carnivorous, feeding on other arthropods and, in some cases, other moths!" Ultimately, the extreme biodiversity of North Carolina means that the state has upwards of 3,000 moth species; in contrast, the same area has only about 175 butterfly species. "And there is still *so much* to learn about them," Merrill said.

According to estimates by the Smithsonian Institution, there are 160,000 moth species and 17,500 butterfly species worldwide.

We had settled on a side porch outside the recently restored Old Orchard General Store, in the tiny town of Lansing, North Carolina, on the banks of the New River. The addition of a pizza parlor, a cidery, and a music venue have brought the old town to life again. Merrill and I have common friends in the owners of this establishment, so we were instantly at home.

As we talked, Merrill would call out the names of birds that we heard singing around us. He started out as a birder, he said, but his work over the years has carried him into many contexts. He once did a stint recording native songbirds for the North Carolina Museum of Natural Sciences—the largest such museum in the Southeast. In 1983, he spent a year conducting a meticulous inventory of natural heritage areas for The Nature Conservancy along the Suwannee River in Florida, including the Ichetucknee Springs Park, where Donna and I frolicked with the river otters in May. He worked for a time with David Beadle, the distinguished coauthor of the *Peterson Field Guide to Moths of Southeastern North America*—the most recent reference book on the subject. We both brought our personal copies of the field guide to this first meeting and laughed at our mutual nerdiness. Merrill is also a longtime friend of Wilson Baker, the Florida biologist who helped preserve the rare trout lilies in Georgia. Wilson had told me about Merrill in February when I asked him about moth-ers I might interview.

Merrill was raised in northeastern North Carolina, in Roanoke Rapids, where his father ran a jewelry store. The historic town is known for the Roanoke River, which offers anglers excellent rockfish bass fishing on the shoals at the fall line on the edge of town. In surrounding Halifax County, where the Lynch family had a farm, the landscape is mostly given to agriculture (cotton, peanuts, tobacco, corn, and vegetables), though there are also broad expanses of cypress-tupelo swamps and bottomland hardwood forests with ample wildlife, including bears, bobcats, pheasants, neotropical songbirds, and wild turkeys. Merrill's dad was an avid quail hunter, but his son preferred fishing and birdwatching. You can still find young Merrill's reports online, typewritten and submitted to *Chat*, a publication of the Carolina Bird Club. Merrill was a stalwart contributor to the statewide Christmas bird count for his home community. He roamed the woods and fields observing and learning. "You can't love something if you don't know it," Merrill said.

He left home for NC State University, where he studied zoology and then took on various fieldwork projects. New rural and suburban developments that dictated the radical conversion of wetlands were overtaking many communities in Eastern North Carolina. Merrill got involved in conservation activities with the state's Natural Heritage Program, which would eventually lead him to The Nature Conservancy. More recently, he collaborated for a time with Bo Sullivan, a professor of biochemistry and a preeminent lepidopterist at Duke University, to help find and identify new moth and butterfly species in the field. "There were not that many scientists studying moths, and DNA analysis was revealing a number of undescribed species," Merrill said. "Without even having to leave my yard, I was able to contribute to the growing knowledge of moths, which, as a lifelong lover of natural history, opened up another world for me."

The North American Moth Photographers Group, hosted by the comprehensive collectors at Mississippi State University's Entomological Museum, has suggested that the total number of North American moth species identified by researchers with the help of citizen scientists may eventually reach a figure between 13,000 and 15,000, including many micromoths, some no larger than a pinhead. "Perhaps 60 to 70 percent of these will be found in the eastern half of the continent," the experts estimate.

That most moths are nocturnal is part of the historic challenge to species discovery and identification. I think of Larry Davenport, a professor of biology and environmental sciences in Alabama, who told me how he nearly drowned twice and lost two expensive cameras trying to determine the species of the elusive moth that pollinates the rare Cahaba lily—a plant that only blooms on slick rocks in rushing river rapids in three states of the South. Nocturnal creatures require nocturnal researchers, and it goes without saying that some folks are better suited for late night forays in the field than others.

The day we met, Merrill was on a daytime assignment to search for the Appalachian grizzled skipper (*Pyrgus centaureae wyandot*), a rare butterfly that has nearly disappeared. En route to a potential habitat, he took time to talk to me. I showed him pictures of some of the moth species that Donna and I had already photographed around my cabin and beyond. I was familiar with a few of the big, charismatic moths that many people can name, among them the gorgeous green luna moth (*Actias luna*), which, in Little Switzerland (the nearest village to my cabin), tends to fly into the local hotel lobby and alight on lamp shades come May.

Several years ago, Donna photographed a large and colorful Polyphemus moth (*Antheraea polyphemus*) that had landed on the passenger seat of my car when I left the windows open one evening at Wildacres, the Little Switzerland retreat center where we host an annual writing workshop in late summer. I'd also gotten reports of an imperial moth (*Eacles imperialis*) photographed there by a writer friend this season. The lodge lights at Wildacres stay on late into the evening, and the moths come in and tend to settle on the wood paneling in the hallways and stairwells. This already magical place atop a mountain knob in sight of Mount Mitchell thus becomes a live moth gallery in summer.

I told Merrill about seeing the dazzling hummingbird clearwing moths (*Hemaris thysbe*) in the daytime in my flower beds, both at the mountain cabin and at my condo, in Carrboro. "It looks like a miniature flying lobster," I said. Merrill laughed.

Elaborate species such as these, large and eye-popping, are the ones that tend to get people interested in moths, he said.

We talked about catalpa trees (*Catalpa bignonioides*), which are endemic to the Southeast. I told Merrill how my grandfather planted one close to his pond to have the caterpillars, known as catalpa worms, available as fish bait. "Yes; catalpa or catawba, as they are sometimes called here," Merrill said. "They were planted at old home places all across the South." The catalpa sphinx moth (*Ceratomia catalpae*) will usually have two broods of offspring each season, according to the Peterson guide.

"Of course, you know some moths are considered pests," Merrill said. "Every crop plant—soybeans, cotton, corn—has a species of moth that can cause problems. They come in with food shipments worldwide. There's always a caterpillar hidden in the box somewhere."

He explained how moths have cleverly evolved to deter natural predators. "Some moths make ultrasonic sounds to confuse bats," Merrill said. "They have bat-jamming sonar. Other species like the great tiger moth (*Arctia caja*), which is found near here at Elk Knob, advertise themselves with outrageously bright colors and bold patterns. That's a scare tactic to startle birds and to indicate that they are poisonous to eat."

"Some moths in the tiger family look to me like a bad outfit from the 1970s," I said. "Others look like stained glass." Merrill nodded.

"Clearwing borers mimic wasps to throw off predators," he said. "They really do look like wasps." These borers are not drawn to light, so the only way to trap them for observation is to use pheromones (secreted scents). Merrill said he was driving one day in his truck with the windows down, carrying some pheromone traps on the seat beside him. "The scent of the traps was so powerful that it attracted clearwings into the truck in the time it took for a stoplight to change from red to green."

Some moths, such as the underwings (various species in the genus *Catocala*), operate in total camouflage, blending in with tree bark or lichen, as do the caterpillars that precede them in the life cycle, Merrill said. Some moth caterpillars have hairs with the capacity to sting would-be predators, including hapless humans.

These clever adaptations to foil predators, however, are no match for human practices that are increasing the threat to all insects. According to an article in a journal published by the National Academy of Sciences, some moth species are in decline from the proliferation of pesticides, radical changes in land use, new light pollution, and climate change, and "the most severe examples of moth declines are from Northern Hemisphere regions of high human-population density and intensive agriculture."

Amid this bad news, Merrill's enthusiasm for moths was contagious. Armed with a bit more knowledge, I was ready to attract another round of creatures to examine and photograph and to contemplate their fate as the essential nighttime vanguard of pollinators that help keep us in groceries.

Anyone can observe moths by putting out a light for them. Since 2012, the last full week of July has been celebrated as National Moth Week. This event was initiated in the United States and now 120 countries observe it, engaging citizen scientists to identify their local moth species and share the data they collect. Hosted first by Friends of the East Brunswick Environmental Commission in New Jersey, the week encompasses both weekends on either end of the week, allowing volunteers to choose their best time to get involved.

Participants are invited to contribute their moth findings to one of the databases hosted by several partner organizations, including iNaturalist, a popular phone app that links to an international database of plants and animals that have been photographed and mapped by location. There's also Project Noah, a software program focused exclusively on wildlife, through which citizen scientists can submit their "spottings." Several other social media programs encourage participants to post their moth photos during the collection period.

This method of exploring insects is a welcome advance from the era of catching butterflies and moths, killing them in a jar with nail polish remover or another toxic chemical, and then mounting them with straight pins on a board to frame and hang on the wall or stow away in a drawer. The new technologies used for moth and butterfly identification are fascinating. According to a piece in *Vox*, in 2018 the Florida Museum of Natural History tried out a new gene sequencing technology on an old, pinned butterfly and identified it as a new species. The finding occurred "some 60 years after it was collected by a local teen," the story said.

Both moths and butterflies are highly fragile creatures covered with colorful scales, and though moths tend to be a bit bulkier and hairier for insulation against nighttime temperatures, they also tend to fade fast as their colorful scales rub off. Some moths only live for a few weeks.

For the kickoff event of National Moth Week in New Jersey, the sponsors usually offer an evening "moth walk" at a local park. Selected trees along the forest path are prepared with a sweet concoction designed to attract moths. The mixture is painted in patches on the tree trunks. Small groups wander from tree to tree aided by minimal lighting, usually flashlights masked with red cellophane, to observe and photograph the moths that arrive to food. This activity is popular with children and beginning moth ers. The sugar bait tends to attract species that are not drawn to light and may be made from such household ingredients as molasses, beer, brown sugar, and a soft banana, according to the Peterson guide.

Another technique for safely attracting moths involves setting up a "light station," as the moth-ers call it. Hang a white fabric sheet on a deck or porch or at the edge of woods, if possible, and let the moths come to you, just as Merrill suggested. Using cotton fabric is best because the weave allows the insects greater purchase than polyester or other man-made fibers do. (One night, I tried a photographer's bounce—a large, smooth, reflective disk that Donna uses to direct additional sunlight when photographing subjects outdoors. It didn't work at all—too slick.)

Once the sheet is secured, you hang a bright white light or a black (ultraviolet) light, or both, so that the light shines strongly on the fabric. Merrill says a UV bulb with a wavelength of 365 nanometers is preferred, and an LED light is preferable over a fluorescent source.

While hunting for owls in Highland County, Virginia, we met the master moth-er and moth photographer, Stephen Rannels, from Hershey, Pennsylvania. Steve takes high-resolution photographs of moths that come to his portable light station and then creates collages of the images to share with his local hosts or to sell as posters at conservation fundraisers. Steve showed me the setup he uses on his travels. He brings two heavy-duty clip-on light fixtures—the inexpensive kind that house painters use but with the aluminum shades removed. He clips the two lights in the center at the top of the sheet, which he suspends from an eave or a wall, out of the wind—ideally with both sides of the sheet available for the moths to land on. Steve clips his sheet to a rope (clothesline size) across the top. He ties down the bottom corners of the sheet to keep the sheet steady if the wind picks up.

Checking for moths every few hours can be fruitful, if not restful, during the night. You can even turn this activity into a sleepover for kids. Some moths stay put, and some fly away during the night. You can check the sheet every few hours to see who has come to visit. In my most recent moth night adventure, however, I slept through and rose at my usual 6 a.m. and still found several species new to me clinging to the sheet in morning light.

We generally photograph the moths with a phone in white light—not using a flash. We try to avoid shadows and get up very close to the subjects, especially if the moths are quite small. Using your smart phone to focus in on the tiniest bugs may hold unexpected surprises. You won't believe the intricate patterns! In other words, don't be tempted to focus only on the big creatures that land on your sheet.

I then use these photos to attempt identification—either online, with a printed field guide, or both. Some moths are much smaller than a fingernail, as Merrill put it, but my iPhone does an excellent job of enlarging the critter to reveal diverse colors and patterns. I'm still learning, but seeing what turns up is fun—a kind of night fishing, I guess you'd say—and the process is something new to share with overnight guests, especially children.

As I have learned from reading about moth hunts in the forest, many species will fly away when suddenly hit by a roaming white flashlight. So even when you are inspecting your "catch" on the cotton sheet, proceed with quiet and respect. Many moths can hear, so if you come toward a moth perched on tree bark or a shrub or your sheet hanging outside, you want to approach quietly, according to John Himmelman, in his book *Discovering Moths*.

Once you have taken all the photos you wish, you might want to try looking at the visiting moths with a headlamp. Moths have eyeshine, like frogs, spiders, and the other creatures mentioned in the frog chapter.

Retrieve your sheet in the morning by gently shaking the moths loose and leaving them to fly away at will. Most will usually seek their normal daytime habitat, hidden away on tree trunks or in grasses or leaf litter. If you leave the sheet out too long, birds and other predators will consider that you have laid out a buffet for them.

After four sessions of collecting moths over the course of more than a year, I was bowled over by the beauty of these fragile creatures from the woods around my cabin. Moths in the Common Geometer family, for example, rest with their wings spread and appear as if they are wearing immaculate, multicolored capes. If only we could print their designs on fabric and wear their extravagant patterns on our shoulders! Brocades, Daggers, Flannels, Glyphs, Groundlings, Metalmarks, Needleminers, Owlets, Sallows, Twirlers, Underwings, Wainscots—these are some of the poetic names of moth families that you may discover in the southeastern United States. The names of individual species are also intriguing, as you will find in a guidebook.

In his paintings of women in elaborate dressing gowns, the artist Gustav Klimt had nothing on moths. One of the visitors that came to my sheet in the Blue Ridge put me in mind of a Georgia O'Keeffe painting of the red hills of New Mexico. Another looked like a Navajo blanket. Other moths reminded me by turns of the great European movements in painting— French impressionism, art nouveau, and art deco.

The elegance in the design of moths has not been lost on contemporary artists. Merrill said I should look up John Cody (1925–2016), a man who became known as the "Audubon of Moths" for his faithful and dazzling portraits of the insects. Professionally, Cody was a practicing psychiatrist, after an earlier career as a medical illustrator. By the 1980s, his moth portraits had been discovered by Gloria Vanderbilt, were featured in *Audubon Magazine* and then were exhibited at the American Museum of Natural History in New York.

The professional gardener, painter, and herbalist Deborah Davis, whose studio is tucked in the countryside outside Charlottesville, Virginia, has spent years capturing moths at night and gently placing them in plastic bags in her refrigerator to keep them safe and subdued until morning. By day, she photographs the creatures, releases them, and paints acrylic portraits of every species she has found compelling. Each canvas is thirty by forty inches, enlarging the moths to a scale that ignites a new appreciation of their intricate ornamentation.

I saw Deborah's work in Warm Springs, Virginia, at the Warm Springs Gallery and Café, a venue that specializes in works that celebrate the natural world by photographers, painters, jewelers, potters, and other craft artisans. The gallery is open from April through December, and the natural surroundings of the building—gigantic rhododendrons and shady outdoor dining spots—are as satisfying as the artwork that's curated with the owners' special mission to serve nature. Deborah's paintings helped me to see and to look closer, even at the smallest moths that landed on my sheet.

Butterflies, more universally celebrated in our time than their moth forebears, nevertheless offer a great opportunity for discovery. Who knew that butterflies evolved from moths? As the Florida Museum of Natural History explains it: "About 100 million years ago, a group of trendsetting moths started flying during the day rather than at night, taking advantage of nectar-rich flowers that had co-evolved with bees. This single event led to the evolution of all butterflies."

I did a walkabout on a summer afternoon at the cabin to see if I might spot any diurnal (daytime) moths. I found only one, lurking in the jewelweed beside the raspberry patch. It did not remain still long enough for a photo or identification.

*A sampling of colorful native moth species that were attracted by UV light shining on a bedsheet hung outside a log cabin. The moths were collected in the Blue Ridge Mountains of North Carolina during National Moth Week in July.*

When I looked up, though, butterflies were dropping from the poplars up high like leaves. Tiger and eastern black swallowtails were landing in the flower beds, taking swigs of nectar from the big beds of pink and magenta bee balm I had planted just for them. Others were nursing from the common milkweed that volunteers each year in my garden. In another bed, two painted ladies were taking turns visiting the coneflowers and orange milkweed. Butterflies are hard to miss when they arrive.

I am ashamed to admit that I used to pull up the milkweed volunteers, frustrated by their random appearance and big leathery leaves, but the scent of their blooms, I now know, is a fabulous perfume. Even though I have planted amply for them should they pass through here, we rarely see a monarch butterfly at the cabin anymore. According to the US Fish and Wildlife Service, the population of monarchs in the United States has declined more than 80 percent in recent years.

How can you tell butterflies from moths? Besides their preference for the daylight, butterflies have antennae that are clubbed at the tip, while moths have more threadlike or feathery antennae. Moths make cocoons out of a single strand of silk to pupate, but a butterfly pupa, called a chrysalis, is naked and is simply attached by a silk thread to a twig or leaf until the butterfly's wings are mature enough to open. Most moths lay flat when they land. Some are fan-shaped. Others look like stealth fighter jets. Some don't look like winged creatures at all but rather like lumps of lichen. Some moths have forewings much like a short cape that overhangs larger, hind wings. Butterflies, by contrast, generally perch with their wings folded upright above their narrow bodies.

In my own gardening journey, I have been fortunate to learn about the current challenges to butterfly survival from the distinguished entomologist and author Dr. Jaret Daniels. He is professor and curator of the McGuire Center for Lepidoptera and Biodiversity at the Florida Museum of Natural History in Gainesville.

The McGuire Center houses the world's largest lepidoptera research facility along with one of the largest collections of butterflies and moths on the planet—some 40 million specimens. The Florida Museum of Natural History is a joyful place for children to learn about animals and plants. As Florida's official state nature museum, the facility hosts a raft of lively, multimedia, and hands-on programs indoors and outdoors for all ages. The bookstore has a vast collection of new books and classics on Florida's environmental treasures and tragedies. A playful array of brass frogs clings to one outside wall of the complex. When we visited, an enormous red replica of an arachnid was peeking over the roof at the front of the building, its legs moving in the breeze as we passed under it.

Donna and I met Jaret in 2022 at a dinner before the annual Georgia Native Plant Symposium. He invited us to Gainesville to see his lab and witness some fieldwork with monarch butterflies—a long-running monitoring project on private land that has provided local data on the species every spring since the 1980s. Population numbers of monarchs have vacillated in Florida, Jaret says, and his lab is also looking at trends across four other projects that focus on the nontarget effects of insecticides on monarchs in the United States and Mexico. At the same time, Jaret's lab has been collaborating with colleagues at Duke University to monitor milkweed populations in the region over time. The number and scope of the Daniels Lab's projects are as impressive as they are pragmatic and clever.

More than a decade ago, Jaret began a solo effort—the first of others to come—to save a federally endangered Florida butterfly—the Schaus' swallowtail (*Heraclides aristodemus ponceanus*). The species was down to four individuals and about a minute away from extinction. Jaret brought it back from the edge by painstakingly transplanting eggs from the wild to his lab. In the first round, he hand-raised seventy-five butterflies to adulthood before releasing them back home. The restoration process has continued. As of 2021, there were 1,700 Schaus' swallowtails living in their native South Florida habitat.

Beyond the scientific interventions, Jaret and his team have been developing innovative ways to teach the public the critical roles that bees, butterflies, and moths perform in our ecosystems. As Jaret is fond of telling decisionmakers, "The top seven crops in Florida are pollinator-dependent and responsible for a figure north of a billion dollars in the state's economy."

His team has reached out to college-age audiences, raising awareness and funds for the lab's insect rescue efforts. In partnership with First Magnitude Brewing Company in Gainesville, which we happily visited, the lab has launched a growing collection of craft beers. Most are named for pollinator species or host plants. Some of the beers have been made with yeast collected by swabbing butterflies. The beer makers grow out these yeasts and pick the best floral notes for brewing.

The University of Florida news service gleefully described the project this way: "Every can or bottle includes a fact about the butterfly organism and an explanation of their importance on the label." The first beer, launched in 2016, was called Schaus' Stout, named for the South Florida butterfly that Jaret rescued. As a video on YouTube about the beer project puts it, "Saving species never tasted so good."

In October 2022, the partners launched REIGN (Restore the Reign of the Monarch), a beer that featured facts about the monarch butterfly on the can, including a scannable link that triggered an app with a 3D virtual reality butterfly. More recently, at the museum's annual Pollinator Palooza, held at the beginning of National Pollinator Week in June, the brewery offered three beers. Honey Bee Citrus Blonde Ale is brewed with orange zest and solar-grown honey. So-called solar honey comes from hives outfitted with solar panels and innovative monitoring gear that gives beekeepers a means to keep the hives at a consistent temperature and humidity, allowing the insects to work more efficiently and produce more honey.

Calamintha Hibiscus Wit, a Belgian-style white ale, is made from locally sourced hibiscus (roselle), a plant visited by the extremely rare blue calamintha bee, which is a Florida native that was rediscovered in 2020 and lives in only fifteen sites in Central Florida. The Daniels Lab is studying the bee's range and habits to support its survival.

The third beer, Frosted Elfin Session Hazy IPA, was named for a nonmigratory butterfly that has suffered from habitat loss. The Frosted Elfin once thrived from Florida to Texas and points north and is now MIA in Georgia and several other states.

At the Pollinator Palooza, guests sampled the brews and took home a free native plant that would attract butterflies, moths, and/or bees. "Every landscape counts," as Jaret likes to say. The more plants out there to serve pollinators, the healthier the planet.

We met some of Jaret's lab crew early one April morning in 2023 on the University of Florida campus. Matt Standridge, a lab technician, arrived at work on his unicycle and stowed it away. He joined Kristin Rosetti, a conservation coordinator, who was already checking on the lab's countertop nursery of Miami blues (*Cyclargus thomasi bethunbakeri*) — one of the most critically endangered butterflies in the United States. It's small, blue, and beloved by those who know it. In the wild, the blues live on the beaches of tiny, isolated islands in Key West and Great White Heron National Wildlife Refuges, in the Florida Keys. Climate change and sea level rise are threats to the species in an already harsh environment rife with blistering heat and raging storms.

Here in the lab, the emerging insects are housed in recyclable cups: some paper, some plastic, all coded with numbers or letters and lined up on a long countertop that runs the length of the room on one side. Each cup in the Miami blues group holds a chrysalis attached to the sprig of a host plant. Grow lights keep them warm.

I laughed out loud when Kristin explained that the lab uses Gatorade as its artificial nectar, served up on Q-tips. "They like Fierce Melon flavor best," Jaret told us. The sports drink Gatorade was invented at a lab here at the University of Florida in 1965, and it shows up pinkish on the nubs of the Q-tips in the cups. At the far end of the room, laundry hampers made of netting house larger host plants and mature butterflies in preparation for their release into the wild.

Jaret's team is also raising frosted elfins (now with their own epony-mous beer). This species would normally be living in a pine savannah in northern Florida and is being considered for federal listing as endangered. It feeds on lupine, recognizable in smaller cups along one section of the countertop, where many pupae are maturing.

Jaret introduces Sarah Steele Cabrera. She is a research assistant and PhD candidate who is wearing a Miami blue T-shirt in honor of her charges in the lab. Jaret shakes out a few pupae from a cup into his hand to show us the elfins up close. This captive breeding will allow the team to restore the thriving specimens to their original habitat, where they will be moni-tored regularly.

Some of these monitoring excursions in the field take place during the hottest months of the year. Matt and Kristin later admit that they both welcome and dread the hot field work in a region with sea water that is intensely blue. In the case of the Miami blues, the work takes place on a deserted island in the Keys where the mosquitoes are big enough to ter-rorize even the most devoted entomologist.

On the opposite side of the lab, workstations with small laptops, print-ers, microscopes, and other devices for analysis cover the countertop. The walls are pinned with large, colorful images of the at-risk insects and plants that the team is working with and the gorgeous habitats they regularly visit. High on a rack of shelves near the door are empty sample cans of the beers that have helped to support this teeming butterfly nursery.

In comes Chase Kimmel, a tall, ebullient postdoc who works with cor-porate and private land managers to determine how to protect wildflowers that benefit Florida's insect communities. The Daniels Lab is concerned not just with preserving pristine areas but also with solutions for heavily modi-fied urban environments and roadsides, where routine mowing practices may be ill-timed for pollinators. Chase worked on a recent project with Duke Energy that looked at utility rights-of-way—prime spots where good conservation and restoration practices can prevent the fragmentation of insect and other animal habitats. Chase is also investigating the habitat re-quirements for three species of bees, including the blue calamintha.

We greet Ivone de Bem Oliveira, a new postdoc research associate who has gained critical experience in big-data analysis and agricultural plant breeding from recent graduate work in her native Brazil. Her research here in population genetics and conservation genomics will help guide decisions about appropriate conservation of imperiled insects. A Latinx, Native American woman in STEM, Ivone is keen to mentor young, would-be naturalists in this work.

Outside, we load up the car with equipment, and soon Jaret is driving us away from campus. Donna is in the back seat with Matt and Kristin, and we are off to see the privately owned ranch south of town where the lab has monitored the annual monarch migration for decades. It is a stable site, where a couple of generations from two different Florida ranching families have continued to support the ongoing research by making their land available to the team.

Lately, the Daniels Lab has been going to the site to train drone flyers to conduct plant surveys from the air, experimenting with what can be seen from different elevations and dealing with shadows and other impediments to good imaging. In 2017, Jaret led a project in cooperation with the Florida Department of Transportation (FDOT) to map butterfly weed along roadsides from central to northern Florida as the first step toward developing strategies to reverse the population collapse of monarchs here. The use of drones will enhance this effort.

He continues to work with FDOT to plant milkweed and other host plants for native pollinators along roadsides, particularly in existing retention basins. These plantings revegetate basins, both dry and wet, and are aesthetically pleasing to motorists. Jaret says, "When mowing mimics the disturbance that some native plants need, it improves the habitat for pollinators and other wildlife, including aquatic life—alligators, frogs, damsel, and dragonflies. Monarchs colonize it right away, we've found."

FDOT has agreed to enhance existing plant densities by reducing their vegetative maintenance along medians and rights-of-way. Jaret's 2017 study showed that the back slope of roadways adjacent to scrub and pineland habitats holds the greatest number of milkweed plants already available to pollinators.

The main species that draws the monarchs to this ranch site is sandhills milkweed (*Asclepias humistrata*). It emerges from the soil in early February and produces nectar-rich flowers that eventually become erect seed pods. The cattle that graze the field here don't bother the plants and their milky sap, though gopher turtles sometimes munch on the leaves, Jaret says as he parks the car in a vast field inside the fence.

*For more than three decades, monarch butterflies have been studied on this Florida ranch where they lay eggs on a local species of milkweed. The egg becomes a striped caterpillar (larva), then tucks into a chrysalis (pupa), and finally emerges as a butterfly (adult).*

Matt and Kristin get to work. They are an efficient team. First, Matt measures the distance between individual plants and then examines each leaf for eggs, larvae, pupae, and adult butterflies. Kristin records the data. Matt uses a net to catch the flying monarchs, stowing each one flat in its own envelope until the census is over so that he won't count any butterfly twice. It is thrilling to see this much-publicized orange beauty flying about. Kristin says this has been a great year so far—"crazy good"—for the milkweed species on the ranch. It is not a species that is easy to propagate in the greenhouse, but the population here has increased by about a thousand plants this year, even in this rather harsh habitat.

Jaret explains that he'd rather invite people to enjoy butterflies in the wild and encourage them to attract insects by planting pollinator plants anywhere they can. "Citizens trying to breed monarch butterflies is not the answer, but we do want to stimulate people to do something for the species. You can attract butterflies to your yard, and that has immediate rewards."

The pandemic supercharged sales of plants in Florida in recent years, and Jaret expects that trend to continue. The museum puts on two wild-flower sales each year, which draw large crowds, including people from out of state. The sales raise significant revenue for the organization. "Workers with technical skills in landscaping, plant breeding, and conservation management are in great demand and will continue to be," Jaret adds, speaking of the job markets available to his students and those at the local community college in Gainesville.

With the growing public interest in planting native pollinators and the concern for butterflies that has moved many to action, entomologists at the University of Florida are developing a certification program in wildlife-friendly plants for garden stores, he said.

The use of pesticides on plants sold in garden and big-box stores has become a consumer concern. Plants may be treated with chemicals by growers so that they look their best on the shelf at the point of sale. But that early application of poison can have a detrimental effect on the birds, bees, and butterflies that visit the blooms once the plants come home to the consumer's garden. "The most common chemicals used in nurseries are not terribly toxic," Jaret said, "but there is a danger from toxic drift from the multiple chemicals that are used across a larger nursery setting. There's no way for the public to know how safe a plant is. If a plant looks pristine, with no insect damage, it may be because of efficacious pest control." Labeling plants that are safer for wildlife and insects is the goal of the certification program, which has yet to come to fruition.

In the meantime, Jaret and his colleagues have some solid recommendations for all of us who garden or want to begin gardening. These ideas are adapted from a paper published by Akito Kawahara, an associate curator at the Florida Museum of Natural History, and other scientists:

1. Diversify and simplify your landscape. Let at least a portion of your lawn go wild to attract insects. Leave piles of leaves untouched where insects can overwinter. Cut back on mowing frequency.

2. Plant native plants. They use less water, are adapted to the local climate, require less maintenance, and draw native insects that belong in your habitat.

3. Reduce pesticide and herbicide use. Chemicals can hurt non-target insects and plants while also contaminating ground water. Native plants and insects provide natural pest control.

4. Limit outdoor lighting. Turn off extra lights or put in motion-activated lights. Avoid bug zappers. Light pollution has doubled in recent decades, threatening our biodiversity.

Jaret told his audience at the Georgia Native Plant Symposium to "plant a billboard" in their yards or in pots on an apartment deck or balcony to attract pollinators. That's exactly what I did at the cabin and at the condo: big colorful patches of native flowers to attract bees, butterflies, and moths. Our homeowners association, in Carrboro, has encouraged condo owners to use a patch of the common land that's near each front and back door and dress it up with natives, even if just in pots. At least 80 percent of Americans now live in or near cities, says the US census, so small-scale urban plantings can make a big difference.

We end our visit at the ranch before the day gets too hot. Jaret, Kristin, and Matt will take off tomorrow for a conference in the United Kingdom, where they will meet with colleagues from Europe, Southeast Asia, and Africa to share their research and learn from others who are working in very specialized and applied moth and butterfly research, most of it conservation specific. It will be Kristin's first trip outside the United States, and she is stoked.

When we get back to campus, the three insect advocates leave us with two more recommendations: where to find the best pizza in Gainesville (Satchel's) and where to see a remarkably diverse preserve for bison, wild horses, alligators, and 300 species of birds (Payne's Prairie Preserve State Park). Both were excellent, as advertised.

Jaret's DIY project to plant a billboard is a venture that's particularly important in the not-so-wild but still biodiverse South. His book, *Native Plant Gardening for Birds, Bees & Butterflies*, focuses on the Southeast and offers an encyclopedic list of pollinator plants along with the insects they attract.

I can testify that an ordinary flower bed brimming with native pollinator plants is fabulous in all dimensions: the sweet scents, the sounds of bees, and the sight of flowers and visiting butterflies and moths. You can even include the tastes of edible herbs or add coral honeysuckle vine (*Lonicera sempervirens*), which my grandmother always had in her garden so that we could taste a drop of nectar from a single bloom, just like the hummingbirds do.

Human-dominated spaces, as Jaret calls them, can make a significant impact in sustaining wildlife, including insects, if planted with pollinators. He says, "For this strategy to work, however, we must all participate and understand that the choices we make in our landscapes are relevant." Every person and every landscape matters, especially if we all take on a small part of the restoration work ahead.

# 12

# Forest Bathing

MONTEREY, VIRGINIA

*Absolutely unmixed attention is prayer.*
—Simone Weil

We sit outdoors at a dining table on a covered patio. Before us, the August sun wheels across a field of waving wildflowers: Queen Anne's lace, field thistle, narrow leaf plantain, joe pye weed, and dock, rising among the tall grasses. The breeze from the high mountains to the west is delicious.

Our host and guide, Annette Naber, is giving us the lay of the land: "A few years ago, there was a huge hillside of pink lady slippers up on Allegheny Mountain, north of here," she says. "But the Forest Service clearcut the timber on top and opened the slope to light and runoff—too much for the lady slippers to survive. They changed the microclimate there." Annette still mourns the loss.

Otherwise, this farm in Highland County, Virginia, is a feast of long valley views, unsullied and brilliantly fertile under a big summer sky. Early morning fog tends to rise from the burbling streams that lace the landscape here, but the ghostly apparitions move on quickly as the August sun ascends.

Emerald Mountain Sanctuary, which Annette and her husband, Dan, often share with guests, is a dramatic parcel of fifty-eight rolling acres tucked along the eastern slope of the Alleghany Mountains. The West Virginia state line runs along the far side of the ridge we see before us.

For twenty years Annette and Dan had careers in Washington, DC. Then Dan, in his fifties, took an early retirement, and they bought this land. "This is our playground, our painter's canvas," Annette says. They grow grapes and vegetables, fruit trees and herbs, medicinal plants and edible flowers. They have built a campground on the land for guests, and they host retreats for artists and women's groups. They colead "clear visioning" sessions for individuals, couples, and small businesses who want to move a project from the brainstorming stage to a full concept. Dan also offers yoga and qigong sessions and guides motorcycle tours of the countryside. Annette uses the landscape as a classroom for workshops in journaling, medicinal and edible plant foraging, and forest bathing. We have come for her guidance in the practice of forest bathing.

Christine Frye, a student intern for the summer, has climbed the hill from the guest cabin below and joins us on the porch as our session begins. Christine is an undergraduate at Bennington College in Vermont. She has broad interests: theater, psychology, dance, literature, yoga, and farming. Christine found Emerald Sanctuary through WWOOF (World Wide Opportunities on Organic Farms). Born in Wisconsin and raised in Eads, Tennessee, a suburb of Memphis, Christine began college at Barnard in New York City. A classmate was killed in a park near campus at the end of her first term, freshman year. Then, the pandemic hit. She took a leave of absence during the pandemic school year 2020–21 and returned to Barnard in the fall. She also lost her grandmother and her best friend during that period. It was too much. Christine transferred to the more rural atmosphere of Bennington in the spring of 2022. Working on the farm at Emerald Sanctuary for the summer, she says, has been part of a search for stability and healing. "I chose Virginia because my dad's side of the family all live in the state, and I could visit them during my time on the farm," she added.

Christine's candor sets the tone, and Annette asks Donna and me to introduce ourselves. We confess that we, too, are feeling some tender uncertainties this morning. I have a long-postponed doctor's appointment coming up, and Donna has a diagnostic medical procedure scheduled in a few days that will require general anesthesia. It is surprisingly easy to share this information. Annette, it turns out, has spent many years as a psychotherapist and is skilled at inviting disclosure and creating a space for it that feels safe.

Forest bathing is a deceptively simple-sounding activity. According to the Global Wellness Institute, "Forest bathing and forest therapy [*shinrin-yoku* in Japanese] broadly means taking in, in all of one's senses, the forest atmosphere." But after sampling a few minutes from each of a half-dozen audio books on the topic—and there are many—I had been put off. Perhaps the dramatic urgency of the narratives as performed by the professional readers on those recordings felt over the top to me. By contrast, I recently happened to meet a kindergartener who matter-of-factly explained how his class had experienced forest bathing: "You just go sit in the woods by yourself for ten minutes or until something interesting happens." He liked it, he said.

The Japanese Ministry of Agriculture, Forestry, and Fisheries coined the term "forest bathing" in 1982, during a time in which Japanese business-people were experiencing negative health effects from the demands of the nation's suddenly booming economy. "There were stories of Japanese men working themselves to death," Annette says. Spending an hour or more in a forest environment became a proven practice to ease stress. The idea caught on around the globe and has continued to draw curious newcomers to green spaces.

The evidence-based discipline of forest medicine, established in 2012, has produced many studies of the practice in Japan and elsewhere. Dr. Qing Li, chairman of the Japanese Society for Forest Medicine, was once a stressed-out medical student in Tokyo. "Forest Medicine," he writes, "studies the effects of forest environments on human health and is a new interdisciplinary science that belongs to the categories of environmental and preventive medicine."

Over time, researchers have drilled down to isolate the effects of time spent in the woods—on men, on women, and on structures in the body, including the autonomic nervous system, the immune system, and the endocrine system. A survey of the literature confirms what Qing Li learned as a medical student: get out in the woods and you will be changed.

Study subjects who immersed themselves in the atmosphere of a mature forest or densely green park showed healthy changes in pulse rates, blood pressure, blood glucose, levels of cortisol (a hormone produced by the adrenal glands), and levels of parasympathetic and sympathetic nerve activity. In one major study based on field experiments in twenty-four forests in Japan, the conclusion was that "forest environments can relieve human psychological tension, depression, anger, fatigue, and confusion, and moreover, that they can enhance human psychological vigor."

Annette says she doesn't remember when she first heard about forest bathing, but in her psychotherapy practice in DC, she used "sandplay therapy," a technique in which clients are invited to create a miniature world with small objects, toys, and sand in a tray. The exercise is understood as a way for patients to express themselves without words. Sandplay is said to reveal the unconscious mind and is helpful for patients who are coping with trauma or anxiety. "I have in some ways transferred my work with deep attention and symbols from sandplay therapy into forest bathing," she tells us.

Annette was raised in the Black Forest, near Baden-Baden, a town in southwestern Germany known for its health spas and mineral baths. On Sundays during her childhood, her family went on long walks in the woods. The four children and their parents always searched for wild foods on these excursions and brought home blueberries and chanterelle mushrooms, among other treats. "So, I came with this tool set," she says of her complementary plant foraging skills.

Annette met her first husband in Heidelberg and came with him to the United States. His mother was a self-taught herbalist in the Caribbean, and Annette absorbed much of her mother-in-law's knowledge of tropical plants when she lived in Tobago for some time. Even as a graduate student in the United States, she always found a way to garden. "My landlady allowed me to grow vegetables and a few herbs," Annette says. "Since then, I have always been taking classes to learn more about plants and their medicine. And now I teach to pass on this important knowledge."

When Annette and Dan (her second husband) moved to Highland County, there was no jewelweed growing anywhere on her property until Annette got a very bad case of poison ivy. "The following year, jewelweed showed up right here," she says, pointing to a flower bed. "It said, 'Here I am, love me, make use of me.'" Mountain folk have long used jewelweed (*Impatiens capensis*) as a salve for poison ivy. The fibrous stems are filled with a watery "ointment" that cools the itch and blisters.

Annette, who will soon publish her first book, *Seasons of a Wild Life: The Medicine of Nature Myths, Animals, and Wild Plants*, has connected to the forest and deepened her relationship to plants on this land. "Sometimes plants come to us for a reason, and I trust plant intelligence," she declares. "Our ancestors communicated with plants and the plants gave up their secrets." She cites the late Eliot Cowan (1946–2022), the author of the book *Plant Spirit Medicine*. Cowan studied with Indigenous peoples and came to believe that contemporary human beings living in urban environments have lost many of the sensory intelligences we once had. Cowan argued that human beings are capable of many more ways of knowing beyond the five senses of sight, smell, touch, taste, and hearing. The hunter/gatherer cultures that Cowan studied had an expanded sensory capacity to connect and communicate with plants and to invite their healing properties to work on human illnesses. The folk remedies known by our grandparents and great-grandparents in the early twentieth century held the last vestiges of this awareness, he claimed.

As Cowan told an interviewer with Gaia, a media streaming service, "What I would suggest is that if people are interested in developing better relationships to plants, or for that matter other aspects of the natural world, they could try to make an assumption that a plant is fully alive, fully aware, has feelings and entertain the possibility that its awareness could be more developed than that of ours."

The humility and possibility in Cowan's words is appealing to me. The notion that plants, and particularly trees, are sentient beings is a popular theme running though the current works of Suzanne Simard (*Finding the Mother Tree: Discovering the Wisdom of the Forest*), Doug Tallamy (*Bringing Nature Home*), Peter Wohlleben (*The Hidden Life of Trees*), and Diana Beresford-Kroeger. I am on board with Beresford-Kroeger, who writes in *To Speak for the Trees*: "Every unseen or unlikely connection between the natural world and human survival has assured me that we have very little grasp of all that we depend on for our lives. When we cut down a forest, we only understand a small portion of what we're choosing to destroy."

Amid the almost unspeakable beauty we witness in every direction on this hilltop in northwestern Virginia, the plants and trees beckon us to explore and appreciate them.

"I intuit what's happening on my property," Annette says as we hike toward the forest at the bottom of the field before us. "My skin sense extends to the property, and I am always drawn to a place where something unusual has happened," she says, and invites us to be awake to the same responses in ourselves. We pass by a trio of mature black locust trees long buffeted by storms but still standing as ragged sentinels on the ridge that extends beyond the house Annette and Dan built. Dan has mowed a meandering path for us through the waist-high flower field that winds across the flat part of the meadow and then downhill toward a mixed stand of conifers and hardwoods.

"My grandpa was into birds and mushrooms; my grandmother was into animals," Christine says as we are walking. I tell her how much I also learned from my grandparents early on—for example, the names and seasons of the plants and birds in their gardens. We take our time admiring the wildflowers up close along the way and watch the dramatic movement above of grand cumulous clouds sweeping slowly across an intensely blue sky. As we draw closer to the woods, we can see that the forest floor is somewhat rocky and leaf littered but easy to navigate. Rivers of lush ferns meander through the understory. The well-worn stones that sit above ground are colorful with lichen and embedded minerals—maybe iron and copper, I think.

Our overall assignment is to be intentional, to be as aware as possible of the wooded environment around us. Annette instructed us not to speak once we enter the forest and not to interact with one another. We are to spread out, within sight of one another, but we are to focus only on the immediate landscape and its details, to pick a spot and take it all in with deep breaths and our senses open.

"Once you start looking, your visual field opens up and you can see so much more," Annette says. She invites us to listen to the forest and notice how sounds are different under the canopy of trees. We should also attend to the terrain, how the earth feels underfoot, hard or spongy. We are to look for a single, small object in the forest that speaks to us. We will bring these totems back when we return up the hill to the porch. Annette says two hours in the forest can reverberate for up to a month in our bodies. It already feels good.

I walk into a heavily shaded area, noting the patches of brilliant green moss and several pine seedlings that are about my height. It seems that when the sun comes from behind a cloud, the sound of the droning August insects swells with the light. A slash of red flies through my field of vision when a cardinal, with urgent, flapping wings, abandons a holly nearby. Then there is another cloud and a decrescendo of whirring insects, as if they, too, are catching their breath. Their music soon drones again and begins to feel abrasive. Leaves drop in the breeze. I take slow steps deeper into the woods, looking up and down as I go. There are sturdy oaks, ash, and hickories; sugar maples, black cherries, and pine. American hornbeam is spreading low to the ground. Looking across the forest floor from here is like being in a theater, with different parts of the stage lit by turns as the clouds move overhead.

I find a hickory nut, pick it up, and sniff it. This smell is my childhood, the forest where I grew up outside Atlanta when the city was much less developed. The trees dropped them like hailstones on the roof and around our house. A hickory nut carries a scent like no other, a bit harsh like menthol but earthy, too. I rub its rough, suede skin, which clings in scored segments to the nut inside. The scent transfers to my fingers.

As a child, I spent a lot of time alone in the woods around our house. I did have a sense of trees as beings, watching me from above, more constant than some people in my life. The reliable scent of hickory nuts was a marker of home, as familiar as the sound of a wood thrush or a blue jay. I sit down on a rock. I close my eyes to listen more carefully. Take slow, deep breaths. The wind stirs the canopy. There is good in this wind, I think.

Oxygen levels are usually higher in the forest because of the working trees, Annette had said. She also told us earlier how pines, spruces, cedars, and even oak trees also give us the gift of phytoncides, which the trees produce to protect themselves from insects. Researchers can detect phytoncides in forest air, and humans can smell them, too. Studies suggest that these "antimicrobial volatile organic compounds" stimulate anticancer proteins, also known as natural killer (NK) cells, in humans. They also decrease the hormone levels associated with stress.

I think of Christine, trying to get her bearings in a world jangled by COVID and with so much random violence. I admire her admission of stress and disorientation. As a freshman in college, I was homesick. I had forgotten, until this moment of stillness, how one day, on a hike with friends, I convinced them to help me carry a four-foot-long chunk of a robust cedar tree that had been downed near campus. We hauled it onto the dorm elevator. I dragged it to my room and leaned it in the corner next to my bed so that I could wake up to the scent of cedar every morning. My father, ever the handyman, had put up cedar paneling in the den of our house, and the smell of that sawdust was one of my earliest recollections of a pleasant and unparalleled fragrance. Maybe I needed the phytoncides.

I asked Donna later what happened for her during that first hour in the forest. "The isolation took me a minute, and then I completely forgot about you all," she said. "I found myself near a tree with three conjoined trunks, like sisters, and I was thinking about my family." Donna lost one of her three ebullient sisters to cancer a few years back. "I was thinking how strong those trees were, rooted together." Donna said she loved not taking pictures, for once. "I was trying to feel, to smell, to have my skin and ears respond without an agenda." She focused on the layers of leaves, imagining what was going on underneath the surface. She is a very private person when it comes to her physical health.

"Annette instructed us to be still and to ask the forest what I needed to know and to be open to what the forest told me," Donna continued. "I felt calm and then an ease from stopping and being still. I had been anxious for days about the medical stuff. Now I was aware of something bigger than me. I had no idea where you all were. I was completely immersed—it was not distraction but immersion," she said.

*Under the tree canopy in Highland County, Virginia, in August, forest bathers found bear corn, Usnea lichen, Indian ghost pipe, and blue-green cup fungi.*

After a time, Annette calls us back together. She points out the dead trees around us and reminds us how they serve as habitat for animals and food for fungi. Decay and death are integral to the forest system, she says. Christine asks about the impossibly bold splash of orange fungi on the side of an oak tree that we all noticed coming in. It's a large and healthy spread of chicken of the woods (*Laetiporus sulphureus*), Annette tells us. She says she'll send Dan to harvest this bountiful mushroom later today. She knows how to prepare it for a delicious supper.

Annette also points out bear corn (*Conopholis americana*) growing low out of the ground. It stands on end, cob-shaped with yellow and red kernels. She says it was also called squawroot back in the day, a term now considered offensive. This perennial is a parasite and does not photosynthesize. Bears often eat it, for its laxative effect, when they emerge from hibernation.

Annette shows us a specimen of *Usnea* lichen, a dainty, light green, string-like moss that clings to tree bark and hangs from twigs, almost like a patch of hair. It is known as powerful medicine against strep and staph bacteria, and it was also used on ancient battlefields to staunch the bleeding wounds of soldiers.

We follow her a few steps farther to a delicate outcropping of Indian ghost pipe (*Monotropa uniflora*), which I have seen before in the yard at my cabin in the Blue Ridge. It comes up under the white pines, and its eerie, white, translucent stems are extremely fragile. At the end of the stems are subtle urn shapes that hang downward and look almost like a woman's half-slip with lace at the edge. "That is heavy-duty, deep medicine," Annette says. This native perennial, endangered in some states, is a great source of lore. We all bend down to observe it more closely.

I have read that Indian pipe is parasitic, attaching itself to mycorrhizal fungi in the soil. Though it's in the same family as blueberries, azaleas, and rhododendrons, it's strangely pale and mushroom-like. Trees depend on mycorrhizal fungi for minerals, and the fungi depend on trees for carbohydrate nourishment. The parasitic Indian pipe connects them and takes advantage of both. Most sources I looked up discouraged even the slightest consumption of this plant because of its powerful chemistry. Curiously, it was Emily Dickinson's favorite flower, and its spectral image appeared on the cover of her first published book of poetry, in 1890, only six years before her death, at the age of fifty-six.

Now Annette invites us to do a "tree sit," which means finding a tree that speaks to us and that we can observe, touch, and lean on as we rest at its base. "Feel the energy, consider the life in the tree, what it must be like to live in one place, passing through seasons and years."

Silent again, we each set out to find our tree.

The young locust I pick is not huge. It twists and bends to one side. It has suffered some loss of limbs. It is young, relatively speaking, and has quite a way to go to reach the height of its neighbors. I circle the tree, looking up from every angle. I move my hand over the bumpy trunk. Finally, I sit and try to get comfortable, maintaining contact with the rough bark and the lichen that is growing in the crevices.

I try to imagine the visitors that come by here from time to time: bears, turtles, songbirds, crows, racoons, snakes, deer, flies, moths, people. I imagine life with these other trees as neighbors. I consider the way the light changes across this one spot all day and throughout the year. I picture the coming of leaves and the losing of leaves and broken limbs; the power of wind, rain, ice, and snow. The long nights. I ponder how much of the tree is unseen, underground. I wonder how massive the buried part is and how different the temperature is above and below ground in this moment. Then I am back to the layers at work here: the processes of decomposition, water absorption, and photosynthesis; the ants carrying seeds and sugars and other bits of nourishment on endless trails under the leaves, over the leaves. I imagine the earthworms and beetles stirring. My mind eventually slows down. I could take a nap, but then something bites my arm—an insect—and I shake it off.

"When we climbed back up the steep, mowed path through the field, I felt so much lighter," Donna told me later. "The dread had left me." As we brought our objects from the forest to the table on the porch and described our experiences, I found myself tearing up. It was the physical manifestation of relief. I somehow felt more confident about the medical hurdles ahead of us. We have what we need to get through this, I thought. Annette said that visitors to the land often become emotional. "It seems like they are rediscovering something that they didn't even know they were missing."

My found object was a colorful rock with a light green patina and veins of rusty red. It sits now beside the gate at the cabin, to remind me to pay attention as I enter the stand of enormous elders—the white pines that surround the cabin. I now understand that sitting on my own porch in the Blue Ridge is a kind of forest bathing. It lifts my spirits and always has.

I wrote to Christine a year after our meeting, hoping to get her reflections on the forest bathing experience from some distance. She wrote back immediately. She was in, of all places, Japan. She was heading back to school at Bennington within the week. Once she was in her dorm and preparing for another semester, we visited through Zoom. She looked energized and refreshed, her hair longer and curly. She explained that she'd taken five classes at a school in Osaka and completed a month-long internship teaching English at the junior high and high school level in the city. She spent the summer living and working on organic farms. She picked strawberries and helped to gather bamboo sprouts on a mountainside. She took care of chickens, cultivated vegetables, and spent time getting to know the son of a Hiroshima survivor in his sixties who lectures against nuclear war and teaches organic farming to students as young as kindergarten age.

"I liked learning from him, hearing his perspectives, and farming with him daily," Christine said. "We had a set schedule of morning breakfast with the family at the table—all traditional Japanese meals. Then we would farm, going from plot to plot in gorgeous mountains." She visited the Memorial Museum in Hiroshima and the Shinto shrine Itsukushima-jinja, and she traveled to Nagoya, Kyoto, Ehime, and Beppu.

She spent her last month with a family in Japan's southernmost island, Kyushu. "I really got to feel a part of their family," she said. "We often spent days resting on the porch together. They took me on family trips with their friends. I got to attend dinners and festivals and spent a lot of time with the kids. I had to do a lot of letting go. I couldn't understand everything happening all the time because my Japanese wasn't that good, so I had to learn to just let things happen. But I couldn't have done that if I weren't being taken care of by such wonderful people."

Her host family grew cucumbers, bitter melon, and peppers, which they sold in night markets organized by the community. Christine relished the chance to participate in traditional dances in the evenings. She helped the family capture, skin, and butcher a wild boar. "I feel like my time there was really grounding," she said.

Of our forest bathing experience in Virginia, Christine said it stuck with her. "When I first drove out there to Emerald Sanctuary to start farming with Annette, I remember feeling a little bit scared of the woods, almost like it could rise up and swallow me. And then, as I spent more time, I developed a sense of closeness or acceptance with the land." She paused. "I feel like a lot of people move away from connecting with nature because it is scary—they imagine predators in the woods, like tigers circling—but a big part of reconnecting with self and nature is facing that fear and letting go. Forest bathing was just being there and letting the process take place."

When Annette asked us to pick an object to bring back to the porch, Christine said she was reluctant. "I didn't feel like taking anything. I felt like nothing was calling me to take it. I've always felt like that as a person out in nature. Everything is already there, and it's not for me to take with me." For her, the experience was more about being quiet. "I don't know if it washed over everybody the same way," she said, "but looking at death and the life in the forest was a big connection for me."

Now, with a year and thousands of miles of travel behind her, Christine says, "I'm feeling a lot calmer than I did before, in a way where I don't have to push myself so hard. I don't go for so many things at once. I have learned to let things happen a little more." She says the break from college that COVID provided and her extended leave of absence from school taught her to slow down. "Ever since our forest bathing experience, I am trying to think less and feel more. I'm a very heady person, so being outside in nature is grounding for me."

Christine is still exploring psychology and dance and has added anthropology to her coursework this term. She is interested in other languages and cultures, she says, especially Japan's. "I feel lucky in that I've gotten to explore a lot during my leave and in the summers. I want to use that to my advantage." The structured way of life in Japan also seems to have helped her to flourish.

"I am still really interested in healing and well-being work. I'm interested in discovering what is well-being, and I feel like that happens in so many ways. After college, I would like to go back to Japan for a year and explore different farms or teach English there. And then maybe do something a little more structured. For me that might be working in health care —being a nurse or midwife, if I can."

The years of tumult from COVID affected us all, and forest bathing was a welcome balm, a salve as powerful as any Annette could concoct from the healing plants she grows in her garden. Writing this chapter, a year later, I also realize that my forest bathing experience in Virginia has made the sight of any clearcut forest more painful.

I drive by a new clearing on a familiar road in Durham, North Carolina. The highway is by turns heavily wooded and densely developed with offices —for doctors and veterinarians—and small strip mall shopping centers. The most recent destruction—a few acres strewn with piles of cut trees, browning brush, and stumps—is more than disheartening to me. A bulldozer, bobcat, and excavator now sit idle and dirty, their jobs done, waiting to be hauled to the next site. At the roadside, on the muddy red earth drying in the sun, is a contractor's sign—a banner strung up on crooked poles and jabbed into the dirt. The word "clearing" is part of the business name. Removing a forest is not clearing. It is toxic.

A friend and freelance writer, Lisa Sorg, who reports on environmental issues in North Carolina, had just published a story about how an environmental consultant reckoned that the county surrounding North Carolina's capital city of Raleigh has lost 3 percent of its tree canopy every year in the past decade—more than 11,000 acres of trees. It is ironic because Raleigh calls itself "the City of Oaks." The organizers of First Night lower a copper and steel sculpture of an enormous acorn from on high at the stroke of midnight every New Year's Eve in downtown Raleigh. The consultant recommended that the county immediately plant trees to offset the losses and identified 2.8 million sites where the plantings could go, especially in economically challenged areas of the county, where the canopy is thinnest and the people who live there are the poorest.

A couple of hours in the forest may have lowered my vital signs, but the vision of any treeless, muddy lot now raises my ire and my awareness of how much energy, oxygen, and life go missing when a "clearing" takes place. I can still bring the view of the ridge at Emerald Mountain Sanctuary to mind. It is spectacular, but every day as she sits on her porch, Annette Naber remembers the lady slippers that once grew prolifically nearby. She has created a small garden of endangered native woodland plants. It will be up to younger folks with the affection and self-awareness of Christine Frye to continue the task of healing our losses around the globe.

# 13

## Wood Storks and Roseate Spoonbills

*To go into these places with your binoculars down
and your eyes wide open—it doesn't make it easier,
but you can see what's going on. Our landscapes here,
and across much of this country, are open wounds.
Those open wounds bear our attention.*
—Drew Lanham

The strange layers of human impact that pile up on a landscape are not always obvious to the first-time visitor. As Donna and I began our journey to observe wood storks in South Carolina, I was also reading the powerful memoir *The Home Place, Memoirs of a Colored Man's Love Affair with Nature*, by Drew Lanham, the South Carolina poet and MacArthur Fellow who is also Alumni Distinguished Professor of Wildlife Ecology at Clemson University. Lanham has written extensively about the disruption—both human and environmental —wrought by plantation life in his native state. Depending on the crops, plantation sites drew different birds to South Carolina in the nineteenth century. Bobolinks, who breed in the North and winter in South America, were drawn to feed and wreak havoc in the rice fields of the antebellum South. Lanham has written powerful poems about the seasons of agriculture in his home state and how they affected his enslaved ancestors and the wildlife in the region. Today, climate change is bringing different, exotic birds to the Palmetto State.

Lanham grew up on a 200-acre cattle and produce farm that his father ran in Edgefield County, due north of Augusta, Georgia, and adjacent to the fashionable town of Aiken, South Carolina, where he also lived for a time. As Lanham told the writer Natalie Rose Richardson, in an interview for *Emergence Magazine*, he is always mindful in South Carolina that the "quote-unquote *natural* landscape we are looking at was made by black hands."

In mid-September 2022, Donna and I made our first trip to explore the landscape at Silver Bluff Audubon Center and Sanctuary, a 3,400-acre preserve, once part of an antebellum plantation. The land was donated to the Audubon Society in 1975 by its last owner, a Pennsylvania quail enthusiast, Floyd Starr. Rather than leave the property to his children, Starr wanted the Audubon Society to improve the habitat for quail on this land.

We rented small guest quarters in a horse barn in nearby Aiken, more than once named "one of the best small towns in the South" by *Southern Living* magazine. Aiken is horse country. In the early 1800s, a new railroad line connecting Charleston to Aiken inspired coastal residents to seek a somewhat higher elevation (515 feet above sea level) to escape the heat and mosquitoes. Soon, winters began drawing wealthy industrialists from the North for golf and fox hunting. Developers built stables, polo fields, show rings, racetracks, and training facilities. Fox hunting, flat racing, dressage, jumping, and trail riding continue as part of the annual calendar of events here.

The first morning, we made our way through the sprawling pastures on the western edge of town to reach Audubon's Silver Bluff Sanctuary, in a less developed area near the village of Jackson. Over the years, the sanctuary's small staff and an army of volunteers have worked to restore the pine savannah that was here more than 600 years ago—an area that was also home to the first known Indigenous pottery makers in North America. Initially, the volunteers selectively harvested trees in overcrowded pine groves, conducted regular burns to nourish the understory, and planted new longleaf pines where needed. A friends group grew out of the volunteer pool to help raise funds for the construction of educational facilities, hiking trails, and a visitors' center. Regular public programming at Silver Bluff began in 2002.

I had never seen a wood stork and neither had Donna. Even before we got to the preserve that morning, I had been looking intently at the low, watery spots we drove by, hoping to see a large white bird or two with long, dark legs. The road rose and fell as we headed southwest toward the Savannah River, where Augusta sits on the far shore. The sandy soil of Aiken County, it seemed, carried many creeks, streams, and ponds that flowed through bottoms thick with trees. When the road to the sanctuary turned to dirt, I caught a glimpse of something white through the thickets as shallow water glinted in early sun. I pulled over and jumped out of the car. Donna followed with her camera. The patches of white turned out to be a wetland field of blooming water lilies. I felt foolish.

Wood storks (*Mycteria americana*) are the largest wading birds that breed in South Carolina. They weren't always here for reproductive purposes, however. Historically, wood storks came to South Carolina only for foraging, after completing their nesting rituals farther south. But as nesting colonies began failing in the Everglades and other parts of Florida because of rampant development, the birds were forced to abandon their traditional sites and move farther north to breed and to nest.

According to the US Fish and Wildlife Service (FWS), a steep decline in the bird's range and population took place in the middle of the past century, and by 1984, the wood stork was placed on the nation's endangered species list. The falling-off came from the loss of feeding habitats, particularly in South Florida, where the "manipulation of water levels through levees, canals and floodgates" disrupted the bird's breeding habits, according to an FWS report. The storks began moving north to coastal Georgia and South Carolina, where they found more reliable habitats for feeding, even in artificial impoundments. The wood storks began nesting in black gum and cypress trees, particularly in South Carolina's ACE Basin, so named for the watershed/estuary formed by the Ashepoo, Combahee, and Edisto Rivers, which meet at the Saint Helena Sound, near Beaufort.

Wood storks are dramatic creatures, broad-chested and tall. They have a wingspan of five feet and are primarily white, but their wing and tail feathers are pitch black, with a greenish tinge. Their necks and heads lack feathers and are dark and mottled, as are their long bills. Adults will have some pinkish feathers under their wings when the breeding season commences, along with pink feet, which eventually fade to gray after breeding. Juvenile wood storks have yellow bills and grayish feathers on their necks and heads, which disappear as they mature. It takes three or four years for a wood stork to reach adulthood.

Wood stork chicks depend completely on their parents for the first two months, so the proximity of ample food to a nest site is critical. According to ornithologists at Cornell University, when adequate food is not available throughout the entire nesting season, adult storks will abandon their eggs or chicks, and the young cannot survive without their parents.

Storks rely on the extreme tactile sensitivity of their bills to locate fish, crustaceans, and other prey. They require shallow water to feed this way, and they can fish around the clock—even at night and in opaque water.

They feel for fish with their bills open while stomping their feet up and down to rustle up their prey. Their beaks are not only ultrasensitive but can also snap shut in a fraction of a second to seize a meal. According to the Cornell experts, storks will often feed in a flank, marching forward while stirring up the sediment and using "tactolocation" to grab their sustenance. They team up in diked marsh impoundments, ditches, ponds, and tidal creeks. Low water levels at the right time can create an easy buffet for storks, which is precisely how the Silver Bluff impoundments are managed. Here, the storks feed among other wading birds, including white ibis, great egrets, and great blue herons, but these other species must locate their food by sight, not touch.

Researchers first documented the presence of wood stork nests—eleven in all—in South Carolina in 1981. By 2020, the number of nesting sites in the state, most of them closer to the coast, nearly reached 4,000, according to a survey conducted by the South Carolina Department of Natural Resources. The birds prefer to nest here in the high branches of black gum and cypress trees that are growing in water.

As Christy Hand, a wading bird biologist for the SCDNR, explained in an official press release, "We have this diversity of wetlands where storks can feed. If one type of wetland is not optimal for storks, they have several different options."

When we first arrived at the mudflat feeding ponds that the Audubon staff maintains at Silver Bluff, the site was quiet. An older man, who carried his binoculars on a fancy harness that crisscrossed his back, greeted us. He was heading downhill while we were climbing the embankment to the ponds. It was ten o'clock in the morning, and the September sun was beginning to bear down, igniting the mosquitoes. He told us he'd been out for a while with no birds of interest to report. "It gives us old people something to do, I guess," he said, and shrugged. "Maybe you'll get lucky." He smiled and left.

The pond was long and narrow, surrounded by a mat of thick, mowed grass. There were fresh fire ant mounds to maneuver around and the possibility of snakes, I figured. A great egret sat on the far side of the water, striking a classic pose atop a man-made platform. Then all at once a flock of birds unfamiliar to me flew overhead and landed at the far end of the pond near a white ibis already there, preening. Could they be storks? We tried to identify them with phone photos and a bird app. It turned out they were juvenile ibis—that's how little we knew about what we were looking for. Eventually, we discovered a second pond farther back on the land and parallel to the first, this one with a platform that served as a bird blind. Though the morning was now alive with dragonflies and a few birds, we saw nothing that seemed big enough to be a stork.

It wasn't until much later in the day—after lunch, a rest in the horse barn, and a phone call—that we returned for another try. We had consulted with Kimberly B. Fickling, program director in environmental education at the Ruth Patrick Science Education Center at the University of South Carolina at Aiken. She told us there was yet another pond behind the other two and that this time of year the sanctuary staff generally lowered the water level there first to improve the storks' access to shallow fishing. (A fourth pond, we later learned, was also out there but was in a less accessible, swampy wetland.)

As we drove out to Silver Bluff the second time, the afternoon light was getting buttery. We hiked in, and there they were, standing in the water, a wall of broadleaf cattail behind them—three giants in the shallows amid a prancing duo of snowy egrets and one little blue heron, all diminutive by comparison and busy hunting for an early supper in the black mirror of water.

The three wood storks stood stoic, shoulder to shoulder, solemn as priests in robes. When they moved, it was like slow motion—only a step or two, regal and disinterested. Their luminous white breasts were set off by the black trim of the wing feathers folded at their sides. With binoculars, I could see the detail in their dark, gnarly faces and necks, weathered and wrinkled. They were precisely reflected in the still water in front them. I watched. Donna photographed. They blessed us, and we were reverent. It was thrilling.

After a while, the muted clatter of a train sounded through the woods behind us, its horn blowing only after it had gained some distance. A friendly breeze cooled us. The storks' graceful movements were like tai chi: the birds were responsive only to their own kind. They continued to ignore the smaller birds and us. Theirs was a contemplative dance, like calligraphic figures unfolding in a fluid but foreign language. We stood there a long time, partially hidden among young magnolias overgrown with cat brier and wild scuppernong vines, which gave off a fruity perfume. Yellow, unripe persimmons hung from near-bare limbs, alongside purple beauty berry and prickly sweet gums balls that still harbored food for songbirds. When we finally gathered our gear to leave, we moved slowly, too, wanting to avoid upsetting the tableau before us.

On the way back to the car, we passed beside the first pond we'd visited that morning. A single stork was rambling along the flat, marshy edge of

*In September, wood storks stand like robed priests in a shallow feeding pond at Audubon's Silver Bluff Refuge near the Savannah River in South Carolina.*

the water, limping most awkwardly. The bird's struggle was enhanced by the comical presence of a little gray heron jauntily prancing nearby, making a mockery of the stork's struggle. Donna stopped to shoot video and still photos of the limping bird, aiming through the web of a writing spider that hung in an opening between shrubs. In the images from that perspective, the spider matched the stork in size. We wondered if the loner could fly or if it was facing its demise in the winter ahead, shunned and grounded.

Little did we know at that moment that a much deeper story, with many layers of history and hardship, accounted for the presence of these man-made feeding ponds.

Silver Bluff, so named because of the glinting mica on the high banks of the Savannah River nearby, was an important crossroads for the Indigenous peoples of the region. Before contact with European invaders, the bluff was the site of a significant Native American settlement, with earthworks —mounds, terraces, and other fortifications—that archeologists are still excavating and studying. The bluff—now some thirty feet above the river —was chosen for settlement because it was high enough to avoid flooding during hard rains, offered good breezes above the water, and served as a welcome trading site along the Savannah River, which had long served as a transportation corridor for the Indigenous peoples.

Hernando DeSoto came through in 1540, and it is believed that Spanish explorers, seeing the mica, dug in the area for silver or perhaps iron sulfide, an ingredient used in the manufacture of gunpowder. In the 1740s, an Irish immigrant and entrepreneur, George Galphin, bought the bluff and established a trading post that carried his name. He built a sawmill, a grist mill, livestock pens, docks, and cabins for family and enslaved laborers. In 1775, the botanist/illustrator William Bartram visited Silver Bluff and declared it a "very celebrated place."

Galphin was eventually appointed the commissioner of Indian affairs for the region and is given credit for persuading the Creek Nation not to give aid to the British during the seven years of the Revolutionary War, which ended when British forces were removed from Charleston and Savannah. Fort Galphin, as the trading post was named during the war, had been captured and held for a time by the British.

Silver Bluff Plantation, given to the cultivation of rice and cotton, expanded Galphin's continuing economic success after the Revolution. Though his two official marriages were childless, Galphin fathered nine children, three of them with the daughter of a Coweta warrior, others with enslaved women on his plantation. He was instrumental in the establishment of Silver Bluff Baptist Church—one of the first African American churches in the nation, pastored by David George, most likely the nation's first Black minister. Reverend George used the Bible to teach his parishioners to read and write. Ninety of his parishioners sided with the British during the war.

According to the *New Georgia Encyclopedia*, when George Galphin died, in 1780, he owned 40,000 acres of land across South Carolina and Georgia and 128 enslaved people, his own children among them.

In the run-up to the Civil War, Silver Bluff Plantation and the neighboring Redcliffe Plantation were the property of the South Carolina governor, James Henry Hammond. Though he adopted progressive farming practices, Hammond embodied the Old South, a ruthless white supremacist and politician. He became the wealthiest man in the state in his time but was forced into hiding after his sexual abuse of four of his teenage nieces became public. Similar treatment of the enslaved women and their children on his plantations had also been his practice, as his offensive diaries attest. Hammond eventually emerged from ignominy to win a seat in the United States Senate but resigned at secession. He died in 1864. Members of the Hammond family occupied the Redcliffe property up to the 1970s, and in the 1940s, they planted the massive evergreen conifers that line the roadside today—Asian deodar cedars.

The Redcliffe Plantation buildings and grounds have been preserved as a state historic site, open to the public and offering interpretive tours. Visitors can witness the living conditions of the enslaved laborers alongside the elegance of "the big house," where Hammond reigned supreme and helped to whip up secessionist fervor.

The contemporary stork ponds of Silver Bluff are seven miles from the Redcliffe house, on the site of Hammond's grist mill from the 1800s. As late as the 1920s, the mill was still in service. "The original ponds went through several iterations with different landowners after Hammond," explained Silver Bluff Sanctuary manager Brandon Heitkamp, in a telephone interview. "It was even a swimming hole at one point in the 1950s and '60s. Then there was a juke joint down there, and something bad happened. I think someone got hurt or killed and the club went away."

When Audubon received title to the sanctuary property in 1975, exactly two centuries after Bartram's visit, they found the concrete remnants of Hammond's grist mill and the water control structures somewhat intact. Following the donor's wishes, Audubon staff set out to build a self-sustainable bird habitat on the acreage, at first with the goal of improving the quail population, much like the efforts described earlier in the longleaf pine savannahs outside of Thomasville, Georgia.

Then, in the early 1980s, after the wood stork had been declared an endangered species, the National Audubon Society received a most unexpected phone call from Department of Energy officials at the nearby Savannah River Site (SRS). The DOE wanted to help create new stork habitat at Silver Bluff.

The SRS is a massive federal facility covering more than 300 square miles in South Carolina's Aiken and Barnwell counties. Announced in 1950, the United States government seized the land, and residents were displaced so that the Atomic Energy Commission might construct a top-secret plant to manufacture the key ingredients needed to build a hydrogen bomb. The "bomb plant," as it came to be known during the height of the Cold War, was "the nation's single largest building project since construction of the Panama Canal," according to a series of stories written by Doug Pardue, for Charleston's *Post and Courier* newspaper. At the beginning, the plant was generally welcomed because it promised much-needed jobs in an economically challenged section of the state.

As part of the operation, SRS would build five nuclear reactors on the compound, each at least two and a half miles apart, and they would install unprecedented security to protect activities and retain secrecy. While these reactors have never been involved in an accident comparable to Three Mile Island or Chernobyl, the site harbors tons of radioactive materials that were buried in unlined trenches in the early years. Employees were exposed to cancer-causing agents, resulting in compensatory settlements and some litigation that is still under way. The plant continues to process tritium for current US weapons systems, but the other primary function of SRS staff these days is the remediation of radioactive materials created on the site over a fifty-year period.

"Solid radioactive waste continues to be dumped into unlined ditches and buried," the newspaper story reported, but the low level of radioactivity of most of the waste means that the threat to local groundwater is profoundly diminished. However, disposal of the high-level plutonium waste that remains on-site, with its half-life of tens of thousands of years, is an unresolved problem. In summary, the *Post and Courier* story called SRS "one of the most contaminated places on Earth, a storehouse of the deadliest radioactive materials known."

This "most contaminated" place is also home to the Savannah River Ecology Lab (SREL), the country's first national environmental research park, established in 1972. Put simply, SREL conducts research on the environmental impacts on wildlife, plants, and whole ecosystems that have come from the Department of Energy's activities at the SRS. Over the years, the lab has studied alligators, frogs, insects, and birds exposed to radioactive materials. The control data for these studies is often collected from more pristine ecosystems nearby. As mentioned in the earlier chapters on frogs and alligators, SREL is a highly respected lab that serves as a research unit of the University of Georgia. It conducts basic and applied research, trains undergraduate and graduate students, and invests heavily in environmental education and public outreach.

In the early 1980s, when wood storks were declared endangered, SREL had already been following the influx of the birds from Florida—their genetics, nocturnal feeding habits, regional movements, and vulnerability to mercury. When nuclear reactor L on the SRS campus was scheduled to be restarted in 1985, DOE and SREL personnel realized that the release of cooling water from the reactor into an area known as Steel Creek Delta would raise the water level too high for the storks that had been feeding there for several seasons. New habitat was needed to protect these now federally endangered birds, so SREL and DOE officials contacted Silver Bluff staff, inviting them to consider ways to provide stork habitat on the Audubon property, only eighteen miles distant, and to draw the birds there, away from the Savannah River Site.

The National Audubon Society received federal funding to subdivide Kathwood Lakes—the site of Hammond's old grist mill. Foraging storks had been documented by birdwatchers in the grist mill area as early as 1977, but a successful beaver dam had since caused the lake to fill up with dense marsh. Construction of the Silver Bluff foraging ponds commenced in August 1985. The ponds were stocked with stork favorites—blue gill, sunfish, and bullfrog tadpoles—and that has continued every year. Sterile carp too big for storks to consume were put in to help control aquatic weeds. Each pond was outfitted with cross levees—water control mechanisms that allow Audubon staff to manage the water levels in each pond independently, for optimum foraging by the storks.

On our second visit to Aiken, a year later, we came in August for the annual Corks and Storks fundraiser at the Silver Bluff Visitor Center, a brick ranch house with an enormous chimney at its center and a broad deck for crowd overflow. This building houses offices, classrooms, and a small natural history museum, with specimens of the native wildlife in the area. An old highway marker, issued by the National Council of State Garden Clubs and noting the proximity of the Bartram Trail (the path Bartram took through the South), is mounted on the chimney.

Staff and volunteers had laid out a generous buffet. In one room, long tables were covered with platters of fried chicken, deviled eggs, pimiento cheese sandwiches, meatballs on toothpicks, a locally famous poundcake, and a tempting mosaic of home-baked cookies. Patrons wandered into an adjacent room to choose from an assemblage of donated wines and other beverages. Volunteers handed out souvenir fans to deter the mosquitoes and gnats outside.

Guests—dressed for the hot weather—carried their loaded paper plates to hunker down and forage in earnest at the picnic tables under a new covered pavilion and at smaller, round tables set out in the late afternoon shade of massive pin oaks in the yard.

We joined some local friends who already knew their neighbor, Brandon Heitkamp, the sanctuary manager—a slender, loose-limbed man dressed in pink shorts and a black golf shirt. He came toward our table with a big smile. He reminded me of the young Jerry Lewis. His energy was infectious.

Brandon had told me on the phone earlier that he would be closing the ponds to visitors for the entire week before this evening's event so that he could properly concentrate the pond's buffet of fish, tadpoles, and crayfish and allow the storks to feed undisturbed. He wanted to draw the maximum number of birds for our observation this evening. He explained that the very first efforts to draw storks to these ponds had involved decoys. "The storks could spot them even if they were flying a mile above the landscape," he said. "Now it just takes draining the ponds enough so that they can home in on the mudflats. I counted fifty-eight storks wading out there this morning." The record count here so far has been 525 wood storks, according to Audubon records.

I asked about the possibility of seeing roseate spoonbills, another newcomer to the state.

"I'm hopeful," Brandon said, hedging.

Roseate spoonbills—which are one of Florida's most sought-after birds to watch and photograph—have been coming to South Carolina in increasing numbers, along with the wood storks. Their cotton candy appearance is a photographer's confection. Their dramatic pink coloration comes from the volume of shrimp they consume. Of course, there would be no shrimp in these freshwater ponds so far inland. Nevertheless, spoonbills have been sighted here, but only rarely.

Brandon didn't light for long at our table; he took time enough to finish a chicken leg and half a sandwich. Then he rose and jogged over to the center of the yard to greet everyone in attendance and to introduce Tim Evans, Audubon's statewide forest manager. Evans outlined the ongoing projects at Silver Bluff: three hiking trails, a pollinator meadow, a demonstration forest for wildlife, and the work that's ongoing to manage invasive plants around the ponds. He explained how more and more birds that have historically nested in Florida are coming to South Carolina. "We have documented sixteen species of ducks in the ponds," he said. "In December we are covered in shore birds—twenty-four recorded species." (The coastal waters of South Carolina are at least two hours from here.)

Soon it was time to load up our vehicles and caravan through the sanctuary's pristine longleaf forest to the feeding ponds. I rode with Brandon and our local friends in his eight-passenger van. He told us how he started out in forestry and then got into soil science. As a boy, he thought he wanted to be park ranger—his father is a microbiologist—but Brandon's main goal was "getting to work outdoors every day and drive a cool truck," he said, grinning. Brandon was hired in 2010 to rewrite the management plan for the Silver Bluff Sanctuary, based on a study of the soils, which in turn dictated the proper placement for remedial vegetation. He is proud of his work and admits that he has become an avid birder in the process. The hiking trails on the grounds of the sanctuary provide excellent bird-watching, he said, and may include sightings of the rare Swainson's warbler, the much-coveted painted bunting, and an active bald eagle nest not far from the ponds.

When we arrived at Kathwood, the scene at the feeding ponds was very different from our first visit the year before. Nearly a hundred guests were heading for the ponds, keeping reasonably quiet, but the birds were not so quiet. Their scene was a moveable feast, great egrets and storks turned in every direction and spread out in the shallow water that was dotted with white feathers and chaff from the marsh grass. As the people circled around about half the circumference of the third pond, many of the egrets decided to take to their leave, roosting in a single tree in the distance like white candles on the limbs. The storks were happy to stay put in their various stages of fishing and preening in the shallows. They were magnificent, and the water was a perfect mirror of their black and white bodies. The crowd huddled close, and many folks were leaning precipitously toward the water, trying to take it all in. The serious birders had brought their fancy scopes and were setting up tripods. People passed binoculars around and were enjoying the spectacle and shared experience.

Donna and I traipsed down the long side of the pond, where we had stationed ourselves a year before to watch the three priestly storks. Along the way, I stopped and introduced myself to Sherri Fields, who is Audubon's conservation director in South Carolina. In her career, Fields has been involved for three decades in community-based conservation work, serving with the Environmental Protection Agency in DC and later in Atlanta. She then joined the National Park Service in the Southeast and finally served a stint as laboratory director of the Hollings Marine Lab, in Charleston, for the National Oceanic and Atmospheric Administration.

Born in Knoxville and raised in Chattanooga, she tells me later when we talk by phone that her father, a biologist and water quality specialist, rose through the ranks in the Tennessee Valley Authority. She and her brothers developed a fishing habit with their dad, though Sherri confessed that she more often found herself watching the birds overhead than the bobber at the end of her fishing line. She majored in marine science at Hampton University, in Virginia, and then completed graduate work in public administration at Louisiana State University.

Now at Audubon, Fields is a statewide strategist working with the national office to map critical habitat and areas ripe for restoration in the state. "We're looking at what the data tell us about those most critical areas now and into the future, and where those migration corridors might be especially important to maintain wildlife connectivity." Fields affirmed that South Carolina has seen a lot of change in recent years. I had read about the recent appearance of significant numbers of white pelicans in the large freshwater lakes around Orangeburg. "Birds that were once iconic to Florida are now showing up here," Fields said. She has also heard that roseate spoonbills are nesting in South Carolina now, not just passing through.

We talked about the importance of teaching children about birds in this era of upheaval. Fields said, "I have granddaughters now, and when they come to visit, we're always outside and doing stuff to get the kids involved and engaged. You know, lying on the dock to see how many egrets, great blue herons, or painted buntings you can count. Or to see an osprey fly over and talk about how cool that is. These birds are traveling such great distances, and it's just an amazing thing to think about their stories!" She smiled. "The thing that always gets me is considering how far they came to get here and how important it is that we preserve places like Silver Bluff and the other Audubon Sanctuary, at Beidler Forest in the Four Holes Swamp, so they have a place to be." (Francis Beidler Forest, near Harleyville, South Carolina, is the world's largest old-growth cypress-tupelo swamp, with thousand-year-old trees.)

On the way back to our bed-and-breakfast in Aiken that evening, Donna said she met Alyssa Zebrowski, the coastal stewardship coordinator for Audubon. Tonight was this young woman's first visit to Silver Bluff, and she was in awe of the storks. Donna happened to mention that we had also hoped to see roseate spoonbills.

"They are the birds that got me into this job!" Alyssa said. She convinced Donna that we must see them, and Alyssa was positive we would find them at Hunting Island State Park on the coast.

In the chapter on moths and butterflies, I described the pure darkness we drove into at four in the morning as we headed toward a South Carolina beach. I didn't realize at the time that the absence of any artificial light around us was most likely owing to our proximity to the vast Savannah River Site, that once top-secret reservoir of radioactive waste, which we skirted for some fifty miles as we made our way south and east from Aiken. We had crossed "Atomic Road" (SC Highway 125) each time we'd been out to Silver Bluff from Aiken. Prior to 2006, that highway was restricted. Drivers were required to keep moving through the SRS property and could be subjected to vehicle searches if they stopped. Those 300 square miles of federal property were also a vague childhood memory. When our family drove to the beach at Hunting Island from Atlanta, my parents always had to decide whether to cut a wide swath to the north or south of Augusta to avoid the "bomb plant." My father had created a bomb shelter in our basement, and as a reserve colonel in the Marine Corps who served in the Pacific and visited Hiroshima after the surrender, he took the threat of nuclear war in the 1960s seriously.

Hunting Island State Park was our destination on that drive through the dark, prompted by Alyssa's information. When we stood on the marsh boardwalk across the road from the entrance that morning, marveling at the ping and suck of the pluff mud as the tide began its morning turn, I thought again of ornithologist Drew Lanham's description of open wounds in a landscape—how strange the succession of land uses around Aiken. And here, despite tremendous efforts to curtail coastal erosion, Hunting Island has lost about twenty feet of shoreline per year over the past fifty years.

The maritime forest at Hunting Island is still dense with cabbage palmettos that fill in the understory surrounding the live oaks, yaupon holly, palms, and pines. Sunlight on the greenery heightens its scent, which is intense. The lighthouse, dating from the 1880s, is still in good shape, but the cluster of maybe a dozen modest cabins that once sat above the dunes at one end of the island—always in high demand as rentals—washed away in 2016 along with the dunes, thanks to Hurricane Matthew.

I recognize our advantages as a family to come to Hunting Island to vacation. In the 1960s, that long beach was a sculpture garden of driftwood. Now only "the boneyard" on one end of the island still offers that landscape of ghostly wooden shapes in the sand. I have pictures of the driftwood that my father made before my birth, when the family was stationed at nearby Parris Island for a time. In one shot, my brother is maybe eight years old, striking a pose on the beach in a Sunday suit, his brogan perched on a driftwood limb, the master of his domain. Another photo, made long after my brother was off to college, has me standing in sand, holding a dead rattlesnake, the snake as long as I was tall. On that visit to the island, my father had heroically run over its head several times as it attempted to cross the newly paved road in the state park. I remember the snake was very heavy. I never questioned the "necessity" of killing it. I also never thought about the gift of such a place preserved.

How I loved this island as a kid! The beach was incredibly flat, making it possible to jump on a skim board to ride a long way through the shallows. I begged my parents to buy me a skim board like the one the Beaufort family we visited always brought with them to the beach. Their son Robert was much more adept at chasing down the smooth, round plywood board and planting his feet on top to balance, catching a smooth ride like a surfer, using only three or four inches of water to buoy him. My parents bought me a skim board, but this was the only beach flat enough for it to work.

After walking on that beach, Donna and I determined that the island lagoon might be where the roseate spoonbills could be found. The saltwater lagoon system here has endured many serious ocean breaches and is still known for excellent fishing. We walked out toward the fishing pier at one end of the lagoon, and sure enough, we spotted the unmistakable pink feathers of the birds, however tiny and distant.

We would never have seen such birds here when I was a kid. Like the wood stork, the spoonbills have been coming north only in the past few years. According to one story from Audubon, spoonbills have been seen as far north as Canada and were spotted for the first time in Massachusetts, Michigan, and New Hampshire in 2021. A year earlier, a spoonbill nest was found in a red maple tree in Charleston County, South Carolina. Two successful fledges emerged.

Like the fate of the wood stork, the roseate spoonbill's success here has to do with water levels, available food, and escaping the disruptive environment in Florida, specifically the Florida Bay, the customary habitat of these flying valentines. Key West, which hosted spoonbills in great numbers at the southernmost end of the Florida Bay, now has water levels that have risen faster than the global average—more than four inches since 2000. Spoonbills, which are named for the odd spoon shape of their bills, are also tactolocators, like wood storks. They feel for their food, and the water where they forage must be shallow for them to find it.

Donna grabbed her camera and took off through the scrub toward the far end of the lagoon wearing flip-flops, which made me shudder. Erosion had long ago torn up the cabin road that might have easily taken us closer to the birds. I followed on a parallel trail through the maritime forest, and both of us got closer, but not near enough to get a close-up image of a single bird. Instead, Donna's best shot was ultimately about context, about encroachment. In the background is the neighboring development

*Roseate spoonbills and other water birds feed in the saltwater lagoon on Hunting Island as development and erosion continue on South Carolina's coastline.*

of Fripp Island, a gated community now, dense with mansions and high-rise condos. Once, it was an undeveloped barrier island occasionally used to test the mettle of marines from Parris Island, where my father had done his basic training.

Donna's photo shows how the condos, golf course, and country club now sidle up close to the eroding state park. It suggests plainly how the birds must hunt for postage-stamp habitats and take what they can get.

On that same summer trip to Hunting Island in the 1960s, our family friends asked Dad if he'd like to go on a boat ride over to Fripp, still thick with native vegetation and wildlife. Property was going for about fifty dollars an acre back then, as I remember them saying. My father declined. He'd already killed one snake that day, he said. We laughed.

# 14

# Bugling Elk

*Paradoxically, the more we study a place, the longer
we know a place, the more mysterious it becomes.
The more we respond to experience, the more we
discover there is to respond to.* —Robert Morgan

Phoebe June Carnes was born in 2003. She is a
biology major at the University of North Carolina at Asheville. Her "chosen passion," as she
describes it, is following and photographing the
herd of elk that were successfully reintroduced
in Western North Carolina only two years before
she was born.

On the last day of September 2023, Phoebe
meets us in front of the Oconaluftee Visitor
Center, run by the National Park Service near
Cherokee, North Carolina. The center sits near the southern terminus of
the 469-mile Blue Ridge Parkway and the North Carolina entrance to the
Great Smoky Mountains National Park. The rutting season has begun, and
the parking area is crowded with curious visitors eager to spot elk, which,
after the moose, is the second largest terrestrial mammal in North America.

Phoebe, who was raised about ten miles away in Whittier, wears blue
jean shorts, a National Parks T-shirt, water sandals, and a ball cap from
the Great Smoky Mountains Association that declares, "I'm rooted in the
Smokies." As she strides confidently toward us, smiling, she cradles her
new Canon camera with its hefty lens in both arms. She suggests we head
for the river out back and down the hill from the visitor center and the
neighboring Mountain Farm Museum. It's a warm afternoon and a bit early
in the day, she says, for the elk to begin appearing.

The Oconaluftee River is a swift, silver thread that begins in the national park and merges with the Raven Fork River on the Qualla Boundary. It meanders into the heart of downtown Cherokee, the tribal headquarters of the Eastern Band of the Cherokee Indian (EBCI).

In Annette Saunooke Clapsaddle's novel, *Even as We Breathe*, she described the river in one graceful sentence: "A cool mist rises from the Oconaluftee as if sighing at the rising sun." Clapsaddle happens to be Phoebe's former English teacher at nearby Swain County High School. She introduced us to Phoebe, and this is our first face to face meeting. Donna and I have already studied the impressive portfolio of wildlife images that Phoebe has posted online.

As the afternoon sun drops, a few anglers wearing waders are out in the river. The Oconaluftee along this stretch is a fisherman's dream, densely stocked with mountain trout. Some 300,000 fingerlings are released each year from fisheries by wildlife officials with the EBCI. A belted kingfisher rattles and swoops before us, heading downstream, defending its territory as a pair of crows chase through the treetops.

The Oconaluftee, Phoebe says, is a magnet for the resident elk, who like to cool themselves in the air coming off the whitewater. We can feel the pockets of cool air on our faces as we walk. The elk can easily remain camouflaged in the shadows across the river cast by a canopy of ancient sycamores. We step cautiously along the riverside trail that is woven with tree roots exposed by seasons of flooding. Until Phoebe pointed them out, I hadn't noticed the obvious cuts leading from the woods to the edge of the river every few yards where the elk regularly make their crossings.

Phoebe soon spots a young bull at some distance on the far side of the river. She calls this animal Caesar and says he is working out his hormonal frustration on a tree branch, yanking it up and down and tearing at the leaves. In the rutting season, young bulls that are hoping to attract a cow will rub themselves against small trees, leaving their scent for females to pick up and often destroying the saplings in the process. But, as Phoebe explains, these "satellite bulls" that hover around the margins of the herd must defer to the oldest alpha male, which is most likely to mate with dozens of cows in the season. The timing is tricky. Females are in estrus for only a couple of days at a time over each of two or three months.

Park volunteers, Phoebe among them, counted no less than twenty new calves born this spring at Oconaluftee, she says. Only three new members of the herd were reported at Cataloochee, where another group of elk claims its primary territory, but there may be more that have not yet been seen.

As we soon observe, Phoebe is completely at ease in this bustling arena of visitors and local folks who are wandering beside the river, looking for elk. They stop to listen to her authoritative comments about the herd and soon sidle up closer to ask questions. The roving Park Service volunteers also know her. They speak and nod as we pass. On crowded weekends like this, they are on duty in their yellow safety vests to make sure that people and elk do not get too close to each other. Warning signs are everywhere —along the roadsides, even in the bathroom stalls at the visitor center —urging patrons to keep a minimum distance of fifty yards from the elk. When she is in a viewing area, Phoebe always carries a stack of rack cards to hand out explaining proper behavior around elk.

In 2020, as a high school student, Phoebe landed a summer internship with the National Park Service and began developing her familiarity with certain elk in the herd. Now she can recognize an astonishing number of the family groups in the field, having photographed them diligently and noted the details of color, size, and related offspring, always at a careful distance. That's where the big lens is essential.

"In my time doing photography," she says, "I have noticed that the younger animals will have more rounded, shorter faces. Then, as they get older, their faces get longer and boxier. This is especially easy to see in the cows."

With the males, Phoebe studies their shaggy "beards" or manes as they are sometimes called, which hang below the head on the underside of the animal's thick neck. "Antler size is also a way to estimate age," she says, "but it's a fickle thing, because an elk's antlers are dependent entirely on nutrition—how well and how much they can find to eat. Some very old males may have smaller antlers," she says. Damage to antlers is another identifying trait. Males fight, often locking antlers, and will have chipped or broken points, the scars of battle.

Early on, Phoebe developed a deep connection with the alpha bull that locals referred to as Big B, for "big boy" or "big bull." "A lot of the park rangers knew who he was," she says, "and I have my own little theory that he might have been one of the most photographed elk in the Smokies. If you look up Great Smoky Mountains elk on Google, he is in lots of images," she says. Phoebe would see Big B on a regular basis, his stature so massive that he was easy to tell apart from the others in the Oconaluftee area. When COVID hit, Phoebe went out nearly every day with her camera for most of that year, sometimes with her father.

The original eastern species of elk (*Cervus elaphus canadensis*) that once ranged over most of the continent east of the Mississippi River, from

Louisiana north to Manitoba, is long gone. In 1791, the naturalist and illustrator William Bartram noted in his memoir, *Travels*, that he had not personally encountered any of the buffalo that once thrived in the South. They had already been "affrighted away since the invasion of the Europeans," he wrote. "The buffalo (*Urus*) once so very numerous, is not at this day to be seen in this part of the country; there are but few elks and those only in the Appalachian Mountains."

These original elk had managed to survive the last Ice Age but were no match for the greedy hunting and habitat destruction of European settlers. The species was as tall as five feet at the shoulder. The males weighed half a ton, with antlers that could add another six feet to their full height. According to veteran hiker and contemporary author David Brill, the last eastern elk was felled by a Tennessee hunter's bullet in 1865.

Though a few private citizens and governmental officials tried to restore elk in the region to serve as game for hunters early in the twentieth century, the process failed. In his environmental history of the Black Mountains, the North Carolina historian Timothy Silver recounted that, in 1900, a hunter named George Gordon tried to stock fourteen elk on North Carolina's Hooper Bald, near Robbinsville (home of the Snowbird Cherokee). When Gordon's project went awry, some of his elk were transferred to the Pisgah Preserve, a 90,000-acre tract set aside in 1916 by the US Forest Service on what had been a part of George Vanderbilt's Biltmore property, outside Asheville.

Another small herd was imported from Yellowstone National Park to the Pisgah site in 1917. Eleven elk from this herd were then moved in 1932 to North Carolina's first wildlife preserve. The Mount Mitchell Game Preserve and Wildlife Management Area was located within sight of the state's highest peak, east of Asheville.

New conservation laws created in the 1930s were not particularly popular with local mountain people, and they continued to poach endangered game of all kinds when and where they pleased. With the Depression in full swing, hardscrabble mountaineers had been known to throw a stick of dynamite into a local stream to harvest a large quantity of trout in one earth-shaking blast. Only six of the elk transplants from Pisgah survived near Mount Mitchell, "and they stuck close to the refuge headquarters, where wardens often fed them by hand," Silver writes.

In 2001, a scientifically disciplined and sustainable project was undertaken to reintroduce elk to Western North Carolina, but the animals that were transplanted to the Smoky Mountains are a different subspecies from the original eastern elk. The newcomers, *Cervus elaphus manitobensis*, are also known as wapiti, a Shawnee word that means "white rump." Wapiti are distinguished not only by their snowy rear ends but also by the bulls' unusual mating call, which is referred to as bugling because of its musicality. In contrast, the male Eurasian elk found in Scandinavia, Eastern Europe, and China make a roaring sound at mating time.

Wapiti were successfully introduced to the Smoky Mountains National Park by way of Kentucky, where they were first established in the Land Between the Lakes after being moved there from Manitoba, Canada. The first twenty-five elk bound for Western North Carolina were hauled from Kentucky to federal park land in the remote Cataloochee Valley in the spring of 2001. They were kept in a three-acre pen for two months, while they acclimated to their new homeland. A thousand observers at a careful distance from the holding pen watched the quiet release of the animals on closed-circuit television. Another twenty-seven elk, transported from western Canada, were introduced in January 2002. Now, Cataloochee is advertised as a prime spot in summer and fall to see the elk, but in recent years, more elk have been showing up along the Oconaluftee River at Cherokee. Park Service officials estimate the North Carolina herd at about 250 individuals, a number similar to the herd in Virginia. Tennessee is estimated to have as many as 450, and there, limited hunting is permitted.

A year before our meeting with Phoebe, Donna and I decided on impulse to visit Cataloochee, where the elk were first introduced. We took exit 20 off I-40 East and followed US Highway 276 to connect to the old turnpike road that leads into the Cataloochee Valley. According to the National Park Service, Cataloochee is a name taken from a Cherokee phrase that means "wave upon wave of mountains." The oval-shaped valley is surrounded by steep mountains. Here, Indigenous people passed through to fish and hunt but did not settle.

Europeans moved in toward the middle of the nineteenth century and used the land to establish orchards and raise crops and livestock. Soon, a grist mill and general store were added to the community, as more than a thousand souls took up residence. Some settlers constructed sturdy frame houses instead of cabins, and they built churches, schools, and a post office. A few original buildings still survive from the controversial disbanding of the settlement to create the Great Smoky Mountains National Park.

The ten-mile drive into the valley is slow, steep, and winding, with three miles of blind hairpin turns on a narrow, unpaved stretch through dense forest. Meeting an oncoming vehicle could be abrupt as we made our way over Cove Creek Gap, but random turnouts allowed passage without our having to run completely off the road when another car appeared.

It was a spectacular Sunday afternoon. Our impulse to see the valley, we soon discovered, was shared by more than a thousand other people, many already parked on either side of the dirt road that dead-ends at the far side of Cataloochee Valley.

At one spot about halfway in, it seemed we had arrived at a family reunion that might have taken place a century before. Among the folks sticking their bare feet into the stream, setting up picnics on ground cloths, and playing tag in the fields was a large gathering of people dressed in the simple clothes of an earlier period—visiting Amish or Mennonites, I figured.

We made our way haltingly to the terminus of the valley road, still negotiating the dusty choke of slow-moving cars and pickups, some with passengers riding in lawn chairs in the truck beds. We found a ranger stationed at the last parking area in the middle of a broad, grassy field. The view was a glorious bowl of color. We asked the ranger if she'd spotted any elk that day.

"They are back there in the woods, I'm sure," the ranger said with some exasperation, strands of hair around her face wet with perspiration and her uniform coated in a thin film of road dust. "But the elk will not show themselves with this many people around." She shook her head and looked across the field. "This is the biggest crowd I have ever seen in here, and I've been working this park for a few years."

We thanked her and turned around. As we resumed our place in the parade of stop-and-go traffic, now heading out of the valley, I thought I heard a strange, high-pitched wail. Our car windows were already rolled down. We had stopped anyway, so I cut the engine for a minute and listened hard. I heard the voices of people coming around one of the preserved churches, the laughter from a group playing frisbee, a growling truck behind us with a radio blaring. These human sounds were nothing like what I thought I had heard coming across the field to our left.

I had purposely not yet listened to elk bugling—plenty of recordings are on the internet—so I couldn't be sure of what I heard. I could wait. I wanted to hear the real thing. The valley was feeling claustrophobic. It took a long time to play our part in the gas-powered parade to get out of the gorgeous, if mobbed, old-home place that afternoon.

Big B, the alpha bull, was very old when Phoebe began studying the herd in high school, and now she wistfully describes the bull as "benevolent. Not that he wouldn't put up a fight, but he was, I guess you'd say, less aggressive. He didn't need to be aggressive, because the ladies loved him anyway." Phoebe smiles. When Big B died a year ago last October, the park biologist thought he might have been eighteen years old, which is nearly a decade beyond the usual life expectancy for elk. Phoebe says she doesn't know the cause of his death, but Big B was deaf and had a swelling on his lower belly. "Some of the biologists I talked to thought he might have a hernia or an infection," she says. "They were hoping in the winter to go in there and get it fixed for him. But he didn't make it."

Big B died only two weeks after Phoebe's final encounter with him, "when I got some of the best photos of him I'd ever gotten," she tells us. "I'll always remember it because it was just me and my dad out there with him. There weren't a whole lot of other people out that day, which is such a rarity in the national park."

Phoebe credits her father with sparking her interest in wildlife. Living right up against the Smokies gave the family plenty of opportunities to camp and hike when she was a child. She has twin siblings—Jesse and Sophie—who are two years younger, but don't seem to share much of their sister's affinity for the outdoors.

"My dad and I would get up on weekends and watch Steve Irwin [of Animal Planet] and BBC Earth and all those animal shows," she tells us. Phoebe has an early memory of going to Cataloochee with her parents and grandparents, and there she saw her first elk. "It bugled, and the odd sound of it made a big impression. I was maybe seven or eight," she says.

Her father, Barry, works for Duke Energy, supervising a team of linemen who service the power grid, often in remote locations where, Phoebe says, he is always watching for wildlife and often comes home with reports. Her mother, Kelly, is a longtime teacher at East Swain Elementary, which may account for her daughter's skill and keen drive to share her scientific knowledge with others. Following her summer with the Park Service, Phoebe began receiving invitations to present her photographs and to talk about her experiences with Smokies wildlife in classrooms and with civic groups.

As we are standing beside the river, an unearthly peal of high-pitched sound issues out of the woods on the far side. To me, it is like a woodwind player warming up, with a range of notes that run from bassoon to oboe to piccolo. "Sounds like Gustavo," says Phoebe, and she explains that this bull is in the running for the alpha spot in the herd. "He's probably trying to round up his ladies and their calves," she says. The humans around us are looking perplexed, maybe a little panicked, turning in every direction, suddenly excited by the bugling, which confirms the presence of a bull nearby. Gustavo's bugle is yet a bit awkward, Phoebe says, which is how she knows him. "The younger ones sometimes struggle to get the notes out," she says. A group of UNC-Asheville students dubbed the bull Gustavo when Phoebe brought them here on a recent field trip.

Bugling involves two separate sounds that elk can make simultaneously. "You have a high-pitched whistle from the nose and then a very low, guttural growl, made deep in the throat, that is harder for humans to hear," she says. Bugling has two purposes: to attract females and to claim territory by warding off other males. Phoebe suggests that Gustavo must be on the move. He is letting his group of females, or his harem, as these groups often are called, know he's moving while also telling everyone else in the area—young Caesar among them—to back off. "That would be my guess," Phoebe says.

A few minutes pass, and we walk farther downstream to a fork in the river, where a fly fisherman is casting a line toward a deep pool in the sunshine. Phoebe tells us that she has seen male elk get into the water up to their necks at this spot when it is hot outside. "They are actually very good swimmers, but you wouldn't think so with those slender legs," she says. Phoebe has captured dramatic photos of bulls swimming and clashing in a fight in the river.

We turn back toward the visitor center. Phoebe runs into a young woman, an aspiring photographer named Ashley, who recently sought her advice on lenses. Ashley lives across the state line in Tennessee, and her parents have brought her here this afternoon in hopes of getting some shots of the elk. As they chat, Donna and I notice that several chickens from the nearby Mountain Farm Museum have come down toward the water to peck at the weeds and insects in the underbrush. It is getting to be a circus here on the waterside.

Now, two female elk come into view across the river—a yearling and a calf that has nearly lost its spots. Newborn elk have built-in camouflage much like that of white-tailed deer. This calf was probably born in May or June, Phoebe says, catching up with us. The two elk approach the water, and soon, there must be a clearing of humans to give room for these females to cross over and move into the sunlight in the meadow beyond.

Then Gustavo appears in the bush and bugles again. He's so much closer, I can hear the air move in his nostrils after he drops his heavy antlers forward and then raises his head back up to issue the call, a sound that begins as a conch shell being blown and then rises to a pleading whistle. My metaphors are lame, however. Nothing else I've ever heard sounds quite like elk bugling.

Dozens of people are now moving along the bank, pulling out cell phones, shooting video and pictures. Gustavo steps gracefully into the

*Behind the visitors center at Great Smoky Mountains National Park, an elk nicknamed Gustavo crosses the Oconaluftee River to catch up with female family members who marched across the water as a crowd of human observers carefully backed away.*

water on the far side and moves toward the middle of the river. He turns his head and Phoebe says he is "snaking"—a term that means he is trying to get the two females to turn back, away from the crowd, but they continue through the opening that the observers have made. The two females are soon stepping into bright sunshine with the chickens dashing ahead of them toward the field.

Gustavo continues a few slow steps downstream and then crosses. He is a massive animal, completely unimpressed by his gawking audience. He steps into the sunlight and prances slowly down the lawn, antlers held high, toward a grazing area that is the size of several football fields. This field is visible from the parkway into the Smokies, and soon a traffic jam will ensue as the elk take their places in the larger group already assembled out there.

The herd makeup is more complex than people generally recognize, Phoebe explains. "The alpha bull stands out, but you also have a matriarch in these groups. When the bull leaves a gathering of the herd, you will still have an elderly female who knows where to find food. During certain times of the year and certain times of the day, you may have multiple matriarchs in a herd. It's pretty easy to distinguish them. You can look for a female that seems like the oldest. But you can also watch for one of the older females to start going in a different direction, and the others will turn and start following her. A lot of people don't give the cows credit, but they're pretty amazing."

Phoebe is well aware of the human tendency to anthropomorphize wild animals by giving them names and assigning personality traits to them. "Anthropomorphize is a bad word in science," she says. "But I do like to give people a more intimate view of these individual animals, and I think that really helps folks connect and respect the animals more and to be inclined to act safely and responsibly around them. I think some of the park biologists would disagree with me on that. They're very much about avoiding names, which I absolutely understand."

Phoebe took a class this past year on communicating scientific principles to a general audience. She already considers herself an informal science educator, and she has worked with lots of children, who relate to stories of animals with names. The secret is, I suspect, that adults do, too. If we live in community with animals and want to consider them part of a community, then naming them can help.

For the Cherokee people, the presence of elk in town is not rare, and some see it as a nuisance. Before this visit, we had come to Cherokee late last fall with my brother to show him around and enjoy the fall colors on the last few miles of the Blue Ridge Parkway. In my opinion, the sourwood trees here are the most beautiful of all on that storied road, and the mountains are relentlessly rugged and multilayered, as seen from the turnouts that face west.

Our approach to Cherokee was from Maggie Valley, an old tourist town that has not lost its 1950s flavor. Retro neon signs on the four-lane highway hawk minigolf and mountain-made souvenirs, low-rate motels, and pancake breakfast specials. Highway 19 eventually narrows and climbs to a flamboyant panorama of the autumn Smokies.

Before we got to Soco Gap, where the Qualla Boundary begins, the first sign of elk was a highway marker painted with a black silhouette of the big-horned mammal against a safety-yellow background. As if the unusual image were not eye catching enough, the DOT added flashing lights around the perimeter of the sign, compelling drivers to be on the lookout. It was hard for me to imagine that elk would try to cross this steep, curving ribbon of pavement that features a solid rock face on one side and sharp drop-offs on the other. Black bear, yes, maybe, but elk?

The descent into Cherokee that day showed us autumn at its most intense. It was as if the trees were lanterns lit from within—sweetgum, sugar maple, yellow birch, buckeye, hickory, and tulip poplar. We stayed on 19, rather than taking the Blue Ridge Parkway, which would be our return route. We ended our descent, coming into the town of Cherokee beside Santa Land, an amusement park opened in 1966 and looking long in the tooth. We passed Harrah's Casino, with its parking lots full of busses. These trappings of commercial gambling have allowed the Cherokee people to build fine museums, fund tuition bills for college-bound youth, underwrite better health care for their children and grandchildren, and provide a stream of other benefits for enrolled members of the tribe. Stewarding their sacred land is a top priority.

We made our way to Big Cove Road, hoping to spot an elk or two. Here, the relatively new and expansive Cherokee Central School sits opposite the Oconaluftee River as it flows into town. A driveway that runs around and behind the handsome campus is called Elk Crossing. That seemed promising. The road has multiple turnouts and passes a park alongside the river. We were enjoying the scenery—rental cabins, house trailers, neighborhoods where folks were grilling burgers and sitting on their porches. Mostly, though, we were watching the river for signs of large animals. But seeing no elk, we turned around and began heading back toward downtown.

We stopped at a turnout where the river appeared to be much deeper than most spots. A rope swing had been installed on a very healthy-sized sycamore that was leaning over a swimming hole, and two-by-fours were nailed into the trunk, forming a ladder that bold swimmers could climb to drop into the water. People were strolling by on the sidewalk, and a man below the sycamore was fishing. Donna asked the fisherman if he'd seen any elk, and he looked at her like she must be kidding. A couple of young women ambled by, and Donna asked them about elk.

"You won't see any elk here! They're all up in the park." They laughed and kept moving. Donna got back in the car, and we drove on. We came around a small curve as the oncoming traffic slowed down. I looked over, and an elk was coming down the concrete sidewalk on our side of the road! His legs were wet and very spindly. He was coming straight toward us. I pulled over; he was so close it seemed risky to lower the window, but Donna did and got photos.

Phoebe confirms that elk do indeed show up in downtown Cherokee and sometimes cause trouble by getting into gardens and corn patches. Because they are not afraid of people, getting the elk to move on is challenging. They also climb the mountain to Soco Gap, where they cross the road day and night, she says. "When females are ready to give birth, they go up high, where their babies will be safe from bear and other predators," she says. So the signs at Soco were quite serious about the hazard of meeting an elk on Highway 19.

In the past couple of years, she tells us, some elk have been the victims of hunters who were seeking their canine teeth, which are made of ivory. These ivories are the remnants of ancient tusks, like those on wild boar, but evolution eventually favored the dominance of antlers on elk. Counting the rings inside the ivories—or any other elk tooth, for that matter—is the most conclusive way to determine an elk's age, Phoebe says.

Last fall, when Big B died, at least five healthy males were in their prime and ready to vie for the alpha spot in the herd. Two of these bulls were found dead, dumped on a farm near Bryson City with their ivories missing.

Now, it seems that the ascendant male is a bull that many locals call Chippy because of the knicks and chips in his antlers. Chippy, Phoebe says, has a most distinctive bugle. He and the other dominant males tend to disappear after the rutting season, and often they will spend the winter at Kituwah, the sacred field and ancient mound that's seven miles outside the town limits of Cherokee on Highway 19. The original Cherokee Mother Town was located there nearly 10,000 years ago.

Before we end our visit with Phoebe, she suggests that we go for a drive to see if we can spot Chippy. When we get to her car in the visitor center lot, she opens the trunk and pulls out an enormous antler. "It was found in my aunt's yard," she says. "It's one that Chippy shed."

The antler is as heavy as an iron bar. I can't quite imagine having two of these appendages attached to my head. "The young bulls struggle," Phoebe says. "They have to learn how to walk through the forest without getting hung up. I don't really see how they get comfortable enough to sleep."

Chippy's antler has seven points, or tines, that are large enough to hold a ring. In hunter's parlance, if a bull has six tines, it is a Royal; seven, an Imperial; and eight, a Monarch. Phoebe says because there are two antlers, they are also referred to by the number of tines on each. According to the picture we showed her of the bull Donna photographed downtown on the sidewalk, he was a six by five—six points on one antler and five on the other.

We head into the Great Smokies park toward Smokemont, and immediately on our right is a cluster of pedestrians and roving volunteers trying to manage chaos. A large group of female elk and calves is in the river under

*The elk herd—mostly females and their young—gathers in fall to lounge at the entrance to the Great Smoky Mountains National Park. The bulls here will bugle to mark their territory and claim their family group.*

the bridge that crosses the Oconaluftee and leads into a parking lot. People are hanging over the railings on both sides of the bridge with cameras. It takes us a while to pass over safely and turn around. By that time, the elk have come up the bank and are crossing the main park road. The yellow-vested volunteers are directing traffic, stopping cars in both directions, and keeping the crowd on the opposite side of the shoulder from the animals. The female elk and calves trot along quickly, making their way single file, back toward the meadow where Gustavo has called a meeting.

We drive beside the Oconaluftee toward Tennessee, glimpsing the meadows on the far side of the river through fallen trees and rhododendron bluffs. In one place, we see a gathering of elk grazing in the sunshine and hayfields beyond. Chippy is there. Phoebe spots him, right off, though his full body height and antlers are partially obscured by the understory between us. We pull over and get out. Soon he bugles and confirms Phoebe's identification. His call is different, four notes in intervals three climb up and one back down.

Several more cars pull over, and Phoebe is at it again, volunteer science educator, explaining Chippy's status to the eager observers. Chippy bugles several more times and finally moves upstream, where a break in the trees makes it possible for us to see him in his fullness.

Phoebe will be back in Asheville for classes on Monday. She will head to New York City in a few weeks for a scientific conference with her biology professor, Dr. R. Graham Reynolds. It will be her first time on an airplane and in the big city. This summer, she will work in the Reynolds Lab at UNC-A on a project studying the conservation and evolutionary genetics of the small boas that are endemic to the Turks and Caicos Islands of the Caribbean. We promise to stay in touch.

On our way back to the Blue Ridge Parkway, Gustavo is holding forth in the giant meadow in front of the Oconaluftee Visitor Center with his harem surrounding him and a satellite bull, probably Caesar, on the sidelines. This group of elk is bigger than Chippy's gang down the way, we think. Traffic is backed up, and children are hanging out of car windows. The males are bugling. Photographers are setting up tripods. Phoebe is among them.

# 15

# Tundra Swans and Snow Geese

POCOSIN LAKES NATIONAL WILDLIFE REFUGE, NORTH CAROLINA

*This wild swan of a world is no hunter's game.*
—Robinson Jeffers

A hard rain pelted the tin roof and thunder rattled the windows. Lightning strobed across the Scuppernong River. It was four o'clock in the morning and unseasonably warm for the middle of winter, even in Eastern North Carolina. I stood at the window, watching the light show from the second floor of the Riverview Lodge—a facility built in the 1920s and now part of the Pocosin Arts School of Fine Craft in the village of Columbia, North Carolina, near the Atlantic coast. Pocosin Arts was hosting a group of twenty nature enthusiasts, who, in two hours, would set out to witness the annual visitation of tundra swans (*Cygnus columbianus*) and snow geese (*Anser caerulescens*) in the Pocosin Lakes National Wildlife Refuge.

I was already nervous. I would be driving most of the guests in an enormous rental van on unpaved roads beside Pungo Lake—a shallow, egg-shaped body of black water. Legend suggests that Pungo is the result of an ancient shower of meteors that hit Earth and carved it out, along with several other remarkably shallow and rounded bodies of water in the area. More likely, the cause was ground fires that burned craters in the peat-rich soil. The 2,800-acre depression that is Pungo then filled up with rainwater.

Our guide for this expedition, Tom Earnhardt, had told us earlier that night over a supper of steamed oysters and boiled shrimp that we were certain to see thousands of swans at Pungo, along with an assortment of migratory ducks, snow geese, red wing blackbirds, and possibly a bear or two. (There are more bears than people in Tyrrell County, where we were spending the night.)

"If you look at Google Earth's night view of the Western Hemisphere, you'll see why the birds come to this spot in North Carolina," Tom said. "The Pocosin Lakes area is probably the largest expanse of true darkness remaining on the East Coast of the United States."

Earnhardt—an attorney, author, naturalist, and longtime host of a state-wide public television show, *Exploring North Carolina*—was prominent among the many conservationists, national environmental organizations, and feisty local farmers who opposed a plan by the US Navy to establish an airfield here. The Navy wanted a dark site where novice jet pilots could practice takeoffs and landings every fifteen minutes around the clock. The proposed airfield would be within three-and-a-half miles of the Pocosin refuge. The birds and planes would be a mutual hazard, opponents argued, and after years of dispute, the Navy finally surrendered the plan in 2008. Today, despite the national publicity surrounding the annual presence of the magnificent swans, human visitors to the refuge remain sparse over the two to three months when the birds are in residence.

It is an hour's drive to Pungo Lake from where we spent the night, and we ply through several stretches of blinding rain before the front moves on and colder air descends. On first inspection, this part of North Carolina's coastal plain seems terribly desolate, especially in winter. Corporate agriculture dominates the long vistas from the main highway where turkey houses and catfish ponds often separate one enormous field from the next. Sometimes a bald eagle or two will be keeping watch on the banks of these ponds for an easy catfish meal. In January, the farm equipment is generally silent. Winter wheat, soybeans, cotton, and sweet potatoes are the rotating crops through the year.

Wild turkey, deer, coyote, and rabbit are plentiful year-round and hide out in the thick scrub along the rivers that lace the landscape. The native scuppernong, a hard-hulled grape, gave its name to the narrow river that meanders north of the Pocosin refuge and widens as it turns toward the much larger Albemarle Sound. More than a decade ago, the red cockaded woodpecker and red wolf—both federally endangered—were reintroduced to the region in hopes of rolling back their near extinction. The two species have persisted. However, the red wolf is hanging by a thread because of the loosening of policies around shooting the animals if they step on private property. Experts at Duke University's Nicholas School of the Environment say that the red wolf is one natural disaster away from eradication in this hurricane-prone wetland. As we have heard time and again in our travels, the red wolf is an example of a species, like alligators and poisonous snakes, that has an important role in the local ecosystem. Here, nutria (*Myocastor coypus*), a semiaquatic rodent, will destroy crops and damage waterways if left unchecked. The red wolf helps to manage and balance the nutria population.

After we pass through the gates to the Pocosin refuge, no one dares take a last sip of cooling coffee from the paper cups we've provided. The ride turns rough. Within minutes, the vehicle is a spattered canvas of sand, slime, and mud. The van waddles over rugged berms and bumps and through pitch-black puddles of uncertain depth. There is no turnaround for miles on the narrow, tree-lined lane, and the van windows fog from the heat of our bodies. We are soon flanked on either side of the vehicle by menacing ditches and impossibly dense scrub. We bounce and lurch forward. The passengers on the third and fourth benches in the back of the van let out inadvertent groans as they are lifted up and dropped back down on their hard seats.

Tom Earnhardt had warned me to keep up my speed and stay a good distance behind his lead vehicle. The heavy van would require consistent forward momentum to avoid lapsing into the muck, tires spinning. Tom was stuck out here once, he said, with no cell phone signal and no tow rope.

I keep glancing at my passengers in the rearview mirror as they cling to the binoculars and cameras around their necks to avoid a flying smack to the chin from their gear. Donna, as usual, is riding shotgun, her camera at the ready.

"As far as we know, tundra swans have been visiting this area in North Carolina since there were people to see them," Tom Earnhardt told me. They navigate their flight from Siberia and Greenland, driven by the increasing cold and decreasing food stores, all the way down the eastern coast of North America. They arrive after the full moon in November and stay for three more moons, until late February or early March. They feed and preen in the shallow waters and spongy fields, and the yearlings among them begin the process of scouting a lifetime mate. The swans don't start breeding, however, until they are home again at the top of the world. It's a roundtrip journey of some 3,700 miles.

According to *National Geographic*, a tundra swan can weigh between eight and twenty-three pounds. It has a wingspan of five-and-a-half feet. Its average life span in the wild is twenty years. Tundra swans, also called whistling swans, are omnivorous, feeding on plants and bivalves in the waters and on agricultural grains ready for harvest in the fields or already spilled from thrashing. Farmers who are allowed to plant crops in the refuge must expect to lose about twenty percent of their production to the feathered visitors, but they say they are so awed by the sight of the swans' arrival that they do not fret. The swans and snow geese, which come a bit later than the swans, are a sign of nature's health, and in any case, they are too numerous for any farmer to chase away, even if given the chance to try.

The Pocosin Lakes Refuge is a maze of roads, mostly unmarked, threading at oblique angles through 111,000 acres of pond pines, gums, cypresses, cedars, and evergreen shrubs. Suddenly, when we come to a dry stretch of road beside a field to our left, Tom's taillights come on. He stops his SUV. Multiple flocks of starlings mixed with redwing blackbirds have just shot up like exploding fireworks from a thicket of trees on the opposite side of the road.

They twist into dark ribbons, by turns wide and narrow, a Mobius strip against the cloud cover. I roll down the window for a better view. The volume of their shrill cacophony proves the number of birds we are seeing —easily a thousand or more, now breaking into separate, geometric bands.

The image reminds me of a toy I had as a child: a magnetic pencil attached by string to a plastic bubble-covered outline of a bald man's face. Under the plastic, a mound of black iron filings could be dragged with the magnetic pencil to create hair and whiskers—mustache, sideburns, and beard—on the man's face. Then you could shake the filings loose and start over.

The starlings seem magnetized by some invisible pencil as they draw themselves across the sky and finally land in the far field. Their dark murmuration stirs us with anticipation. Their white migration mates are surely up ahead.

The trip to a new destination always seems longer in the going than in the return, and so it is with this trip on the pot-holed roads. When we finally come upon a sign for the Pungo Lake Observation Platform and the Charles Kuralt Trail Site, the passengers in the van are giddy. Kuralt, a twentieth-century journalist famous for the human interest reporting he did for the CBS television network, loved this area of his home state. This site is one of several trails that trace his travels and commemorate his stories of the region.

We park near the observation deck and clamber out of the van. Because of the wind direction today, Tom says we will likely find the swans on this side of the lake. As we climb the wooden stairs to the observation deck, he tells us there are maybe ten to fifteen thousand swans in transitory residence throughout the refuge. When we reach the deck and look out, the mass of swans we see before us is staggering.

Our first focus is on the swans closest to us, ranging about in the near distance like hundreds of white buoys unmoored and moving with the wind. Then we discern that yet another massive gathering of birds is on the far side of this broad body of water. Observed without binoculars, the distant birds are a solid, white ribbon stretched across the entire length of the water on the far shore. Then, all at once, the ribbon lifts vertically, and the birds rise into the air, a sight so startlingly abrupt and immense that it raises goosebumps on my arms. The solid white form shatters into discrete chalk marks, swans going in every direction as the birds disperse—first against the backdrop of dark brown trees and on up to the clearing blue of the sky. "Probably an eagle or hawk showed up," Tom says, as all of us on the observation deck shudder in wonder. That was the moment, I think, when I knew I wanted to keep seeing such wonders as this.

After a while, the cold soaks into us. The swans' honking is palpable when the wind slows. Their conversation is somewhat musical, "a sound that suggests a woodwind instrument in its quality," the naturalist Rachel Carson once wrote of these swans. If so, this is the biggest woodwind section I have ever heard, warming up, getting their reeds good and wet and ready for the conductor to arrive.

Carson—the author of the landmark book *Silent Spring* (1961)—successfully challenged the use of DDT (dichloro-diphenyl-trichloroethane) as a pesticide and earned many literary and environmental awards. Much earlier in her career, however, she wrote a series of pamphlets about the National Refuge System. One, published in 1947 about nearby Lake Mattamuskeet, grew out of a visit to this unusual landscape—a trip that she never forgot. Of the winter visitation of swans and geese here, Carson wrote, "At intervals the sound swells as though a sudden excitement had passed through the flock, and at each such increase in the sound a little party of birds takes off from the main flock and moves away to some favored feeding ground."

The English surveyor and naturalist John Lawson was the first person to document the presence of swans in this territory. He set out from Charles Town (later, Charleston) in December 1700 and made his way through what would eventually become South Carolina and North Carolina. It took him fifty-seven days to cover 550 miles. He was a cheerful documentarian, aided by the Indigenous people he met along the way. Of the swans he encountered, he wrote, "We have two sorts: the one we call Trompeters

*From late November through most of February each year, snow geese and tundra swans from Greenland and Siberia come in huge numbers to forage and socialize in the Eastern North Carolina fields and ponds of the Pocosin Lakes National Wildlife Refuge. Photo by Tom Earnhardt.*

because of a sort of trompeting Noise they make. These are the largest sort we have, which come in great Flocks in Winter and stay, commonly, in the fresh Rivers till *February*, that the Spring comes on, when they go to the Lakes to breed. A Cygnet, that is, a last Year's Swan, is accounted a delicate Dish, as indeed it is." The second, smaller species Lawson documents in his book, *A New Voyage to Carolina* (1709), he called "Hoopers."

Tom Earnhardt says in some years he has seen a few trumpeter swans — birds that are "about a third taller" — in the mix with the tundra swans. He also alerted us to small groups of sandhill cranes that may come through with the swans, which we would see on subsequent visits.

We drive to several more viewing sites. At one, we are much closer to the swans because the road runs right beside a canal between ponds. I watch a pair of swans take flight. I imagine what strength it must take to lift a twenty-pound body of wet feathers free of the numbing water. The swans pedal their feet, as if riding a bicycle, as they lift their bodies from the pond's surface. Then, for a moment, they seem to be walking on water, their wide wings pumping gracefully. Once their feet clear the water, their necks are straight shafts, and their bodies seem smaller and streamlined. When they land, they resemble barefoot water skiers, their webbed feet literally serving as brakes, stirring a white wake behind them.

Through binoculars, I can pick out a few young cygnets—the yearlings that John Lawson favored for his dinner. Their feathers are light gray, not yet the brilliant white of their elders, the gray color providing a protective camouflage. If we stayed until sunset, Tom tells us, the mature white birds would take on the pinkish cast of the sky and look like flamingoes or roseate spoonbills.

At yet another pond, one surrounded by more foliage than others, seeing the swans to scale against the cattails, the turtles perched on logs, and the cypress knees creates an even greater appreciation of their size and elegance. A little later, we watch snow geese and swans feeding in a field —their main daytime preoccupation. We stop for our picnic lunch at a spot known as a crossroads for bear, but on a brief hike, we find only their telltale scat, full of corn and cobs from the nearby fields. Tom gives us a little lesson on the value of river cane as habitat for birds and mammals and as a foil for erosion. A native bamboo, it grows easily along the ditches here.

As the skies become cloudless, we see V-shaped wedges of swans moving from land to water and back, set off by the deep blue above them. As we are leaving the refuge, the group in the van talks about wanting to come back. Could it be a two-day excursion next time so that we can see the sunset reflected on the birds? They want to bring friends. Someone declares that this is a bucket-list item you don't just check off and are done with—seeing the swans is as magnetic as the swans' mysterious navigational abilities that bring them back to this spot on their heroic migrations.

Dr. Betsy Bennett, who served for more than twenty-two years as the executive director of the largest nature museum in the South, the North Carolina Museum of Natural Sciences, has taken this tour many times over the years and never grows tired of the spectacle, she explains to the group in the van. "We have to educate children and adults about the value of these wilderness areas. Places like this," she says, "not only protect critical habitat but they mitigate the effects of climate change. I wish more people

(overleaf) *At sunset, the swans and geese take on the pink cast of the sky. Photo by Tom Earnhardt.*

could witness firsthand the wonder and beauty we've seen today. You just never know when someone might be moved to become a better steward and get involved, hopefully with their children in tow." We are all nodding.

That trip to Pungo Lake, in January 2019, was more than a year before the pandemic was declared. Donna and I had already started talking about creating a book full of excursions in natural environments. We were driven by the desire to share marvels like the swans with others. We knew we would come back to Pungo again and again, always a bit on edge because of its remoteness and remembering to check our fuel gauge and other essential supplies before going in. Now, visiting the swans and snow geese is our January habit, like Betsy Bennett's, and we always bring friends along who will be experiencing it for the first time.

On that rainy morning at Pocosin Arts, before we had set out for the refuge, we filled our coffee cups and thermoses and made sandwiches for our picnic with the swans. Tom Earnhardt called us to order and told us that the former US poet laureate Mary Oliver had died three days before, at the age of eighty-three. He wanted us to think of her when we were in the refuge, and he quoted a few lines from her poem "Sometimes." Oliver gathered material for her poems every day by rising early and setting out into the woods or along the seashore, paying fierce attention to the goings-on around her. She advised readers likewise to take the time to attend to and marvel at the world, to value these astonishments, and to spread the word.

That was our simple goal—Donna's and mine—as we began our pilgrimage across this unusually biodiverse region to bear witness to ordinary seasonal events that are accessible to most everyone who has an interest.

What we have found in ourselves, finally, is a deeper reverence for the landscapes and species we have seen and heard: the delicate lightning bugs, the rambling elk, the sentinel frogs, the fragile whooping cranes. Each has been telling us a story of survival against increasingly poor odds. The accounts of successful remediation and renewal for some of these creatures provide hope. And we must also take account of the radical disruptions that our curiosity and fascination can bring to predominantly nonhuman environments, a lesson not learned so well by my generation.

As I was wrapping up this manuscript, I sat down with my first cup of coffee on a chilly February morning in 2024. The sun was not up, but I could hear a wren and a cardinal getting the day started outside, calling to other birds as they came and went from the tall holly bush at the corner of the building.

Half awake, I glanced out the second-floor window right in front of me, and on a tree limb just beyond the glass, my eyes fixed on a dark shape against the barely lightening sky. At first, I thought it was a new squirrel's nest that had somehow been constructed overnight—leaves bunched up on a bare limb.

Then I realized it was animate. It was an owl about the size of a football, perched and watching the parking lot below. This is the time of year I usually hear barred owls calling in the woods around the bioretention pond: *Who, who cooks for you?*

The owl turned her head to look at me and issued a call, high at first and then cascading downward in pitch with a strong vibrato. We stared at each other for another moment. She called again defiantly, and then swooped off her perch, down toward the pavement.

The late Alabama naturalist E. O. Wilson suggested in one of his last books that we, as stewards of the planet, should set aside fully half of the natural world for nonhuman use. Such a concept can only find traction if we help the next generations know better what we are imminently losing and what we still have. The owl at the window was claiming her territory. I am still listening.

# Acknowledgments

To all the named scientists and guides whose personal stories and shared expertise appear in these chapters: thank you for your leadership, time, and knowledge.

Thanks to Jeanette Stokes and Shared Visions Foundation's Jay Miller for their combined assistance with travel expenses related to the research for this book.

I also want to appreciate the naturalists and wildlife advocates behind the scenes who introduced us to new people and destinations in our travels and were always available when I had questions. In Alabama, thanks to Lisa Buck, Pam Bates, Penne Beasley, Greg Screws, Gena Rawdon, and Beth Wicker. In Georgia, thanks to Jennifer Ceska, Carol Denhof, Mincy Moffett, and Atlanta Botanical Garden staff members Laurie Blackmore, Emily Coffey, Ian Sabo, and Loy Xingwen. In Florida, special thanks to the unflappable Gail Fishman. In Tennessee, thanks to Stephen Lyn Bales, Vicki Brumback, Jim Harb, and Laurel Goodrich. In South Carolina, thanks to Jenks Farmer, Tom Hall, Linda Lee, Cole Mitchell, Edward Fort, Nance Lee Sneddon, and Margaret Jenkins Skinner. In North Carolina, thanks to Walter Bennett, Johnny Burleson and Walter Clark, Candice Cobb and Martha McMillan, Winter Gary, Virginia Holman, Ann Huckstep and Carol Misner, Vaughn and Beth Morrison, Elizabeth Palmerton, Feather and Willy Phillips, Bland and Ann Simpson, Alvis Smith and Dennis Williams, Marlene True and Derence Fivehouse, and Janice Worth. In Virginia, thanks to all the Goodalls, especially Pen, Leslie, and Miles. Thanks also to Robin Muira and Samia Serageldin, who published an early version of chapter 15 in the journal *South Writ Large*.

As always, Donna Campbell's participation in this work involves more than making exceptional photographs. She is a thought partner, a researcher, a naturalist, and mighty fine company. Though she is admittedly not too good at navigation, she is always ready for an adventure.

For the design of this book, I am grateful to Kim Bryant and Madge Duffey, and to Bonnie Campbell for her work on the moth composite. Thanks also to photographers Erin Cork, Tom Earnhardt, Terry Priest (www.frfly.com), and Margaret Tyson for sharing excellent images we could not have obtained otherwise.

Now through six books over twenty some years, Debbie McGill has made me seem smarter and more eloquent than I am, thanks to her supreme editing skills. Major appreciation also goes to Executive Editor Lucas Church, Assistant Managing Editor Erin Granville, and copyeditor Alex I. Gergely at UNC Press. Thanks in advance to Dino Battista, Peter Perez, Sonya Bonczek, and Susan Garrett for their good work in promoting UNC Press authors.

I look forward to connecting once more with the bookstores, libraries, book clubs, conservation groups, and literary festivals that have supported my work across the South, especially the On the Same Page Literary Festival in North Carolina's Ashe County, where, over the last fifteen years, I have been allowed to try out various book talks for the first time. And finally, gratitude goes to the generous and supportive Table Rock Writers Workshop community, which continues to provide a rare model of peer learning, advocacy, and unconditional love.

# Bibliography

## Preface

Biewen, John. "The Age of Dominion." *Orion*, Summer 2022. https://orion
 magazine.org/article/age-of-dominion-religion-anthropocene/.

Finney, Carolyn. *Black Faces, White Spaces: Reimagining the Relationship of
 African Americans to the Great Outdoors*. Chapel Hill: University of North
 Carolina Press, 2014.

## Chapter 1. Sandhill Cranes

Alabama Division of Wildlife and Freshwater Fisheries. *Sandhill Crane Study
 Guide*. Accessed February 8, 2024. www.outdooralabama.com/sites/default
 /files/PDF%20documents/Sandhill%20Crane/Sandhill-Crane-Study-Guide
 .pdf.

Atkeson, Thomas Z. *Wheeler: A National Wildlife Refuge*. Conservation in Action
 7. Washington, DC: US Government Printing Office, 1949.

Cornell Lab of Ornithology. "Sandhill Crane Life History." All About Birds.
 Accessed August 10, 2024. www.allaboutbirds.org/guide/Sandhill_Crane
 /lifehistory.

Lundquist, Penny. "The Ripple Effect . . . and Saving Whooping Cranes."
 *iTEAMchicago* (blog). June 8, 2015. https://iteamchicago.wordpress.com
 /category/my-weekly-reader/.

Ress, Thomas V. *Wheeler National Wildlife Refuge*. Images of America.
 Charleston, SC: Arcadia, 2019.

Seamans, Mark E. *Status and Harvests of Sandhill Cranes: Mid-continent, Rocky
 Mountain, Lower Colorado River Valley, and Eastern Populations*. Lakewood,
 CO: US Fish and Wildlife Service, 2021.

Smith, Stephanie. "Whooping Cranes in the Eastern Population—The Journey
 North." International Crane Foundation, March 11, 2022. https://savingcranes
 .org/2022/03/whooping-cranes-in-the-eastern-population-the-journey
 -north/.

Strauss, Emily. "Wheeler National Wildlife Refuge: Home to 30,000 Cranes."
 CritterFacts.com. Accessed February 8, 2024. https://critterfacts.com/wheeler
 -national-wildlife-refuge-home-to-30000-cranes/.

Thompson, Hillary. "Whooping Crane Eastern Population Update—February

2024." International Crane Foundation. Accessed August 13, 2023.
https://savingcranes.org/2024/02/whooping-crane-eastern-population
-update-february-2024/

US Fish and Wildlife Service. "Wheeler National Wildlife Refuge." Accessed
February 8, 2024. www.fws.gov/refuge/wheeler.

US Geological Survey. "CANUS and His 186 Whooping Crane Descendants
(1964–2003)." 2003. https://whoopingcrane.com/cranus-186-whooping-crane
-descendants/.

Walkinshaw, Lawrence H. "Migration of the Sandhill Crane East of the
Mississippi River." *Wilson Bulletin* 72, no. 4 (1960): 358–84. www.jstor.org
/stable/4158860.

## Chapter 2. Bald Eagles

Butler, Harry D. "Eagle Lovers Await Birth of 'Kids' at Lake Guntersville."
Tuscaloosanews.com, February 13, 2014. www.tuscaloosanews.com/story
/news/2014/02/14/eagle-lovers-await-birth-kids/29918835007/.

Cuffia, Ashley. "The Bald Eagle, Creature of Nature and an American Symbol."
*Inside Adams Science Technology and Business* (blog). Library of Congress,
June 27, 2019. https://blogs.loc.gov/inside_adams/2019/06/the-bald-eagle
-creature-of-nature-and-an-american-symbol/.

Elder, Gregory. "Professing Faith: The Eagle Is a Religious Symbol as Well as
a National One." *Redlands (CA) Daily Facts*, July 6, 2018. www.redlandsdaily
facts.com/2018/07/06/professing-faith-the-eagle-is-a-religious-symbol-as
-well-as-a-national-one/.

Erickson, Laura. "About Bald Eagle Nests." Journey North, University of
Wisconsin–Madison Arboretum. Accessed February 8, 2024. https://
journeynorth.org/tm/eagle/NestAbout1.html.

Klein, Christopher. "Did Benjamin Franklin Propose the Turkey as the National
Symbol?" History.com. September 1, 2018. www.history.com/news/did
-benjamin-franklin-propose-the-turkey-as-the-national-symbol.

Love, Shayla. "High in Their Treetop Nests, This Biologist Discovered the Truth
about Eagles." *Washington Post*, August 12, 2016. https://www.washingtonpost
.com/people/shayla-love/.

Millar, Jody Gustitus. "Biological Information on the Bald Eagle (Derived from
64 FR 36453 (July 6, 1999))." Animal Legal and Historical Center, Michigan
State University College of Law. www.animallaw.info/article/biological
-information-bald-eagle.

Oppel, Richard A. "Bald Eagles, Symbol of America, Are Dumping Trash on the
Seattle Suburbs." *New York Times*, April 2, 2019. www.nytimes.com/2019/04
/02/us/bald-eagles-trash-seattle.html.

Rainer, David. "Lake Guntersville State Park's Eagle Awareness Provides Great

Winter Recreation Opportunity and Celebrates Conservation Success Story." Outdoor Alabama.com, Alabama Department of Conservation and Natural Resources. December 29, 2022. www.outdooralabama.com/articles/lake -guntersville-state-parks-eagle-awareness-provides-great-winter-recreation -opportunity-and.

Tapia, Sofie. "Bald Eagles Are Carrying Trash from a Landfill to Seattle Suburbs and People Don't Know What to Do about It." BoredPanda.com, April 9, 2019. www.boredpanda.com/bald-eagles-landfill-garbage-dumping-suburban -yards-seattle/.

Wozniacka, Gosia. "The Dark Side of 'Compostable' Take-Out Containers." Eater, January 15, 2020. www.eater.com/2020/1/15/21065446/compostable-take-out -containers.

## Chapter 3. Dimpled Trout Lilies

Cox, Jim. "Long-Time Red Hills Naturalist Recognized for Contributions." *Tallahassee Democrat*, November 25, 2015. www.tallahassee.com/story /life/2015/11/25/long-time-red-hills-naturalist-recognized-contributions /76318910/.

Dickinson, Emily. "We Like March." In *The Complete Poems of Emily Dickinson*, 130. Boston: Little, Brown, 1927.

Dozier, Patti. "Greenwood Property Sells for $22 Million." *Thomasville (GA) Times-Enterprise*, March 18, 2015. www.timesenterprise.com.

Duncan, Mack. "Fall Line." *New Georgia Encyclopedia*. July 23, 2018. www.georgiaencyclopedia.org/articles/geography-environment/fall-line/.

Florida State Parks. "Steephead Streams." Accessed August 10, 2024. www.florida stateparks.org/learn/steephead-streams.

Johnson, Paul W. "Make Iowa Great Again—Restore Leopold Center Funding." *Des Moines Register*, February 9, 2018. www.desmoinesregister.com/story /opinion/columnists/iowa-view/2018/02/09/make-iowa-great-again-restore -leopold-center-funding/316955002/.

Newman, Joyce. "Trout Lilies: What's in a Name?" *Around the Garden* (blog), New York Botanical Garden, May 7, 2014. www.nybg.org/blogs/plant-talk/2014 /05/around-the-garden-trout-lilies-whats-in-a-name/.

Peach State Archaeological Society. "Hitchiti Indians." Accessed February 14, 2024. https://peachstatearchaeologicalsociety.org/cultural-histories/historic -european-contact-period/hitchiti-indians/.

Smith, Winnie L. "Angus Gholson: A Legend in His Time." *Chattahoochee Twin City News*, January 23, 2014, 4.

Wetherington, Mark. "Wiregrass Georgia." *New Georgia Encyclopedia*. September 28, 2020. www.georgiaencyclopedia.org/articles/geography -environment/wiregrass-georgia/.

Yabor, Erik. "Grady County, Wolf Creek Preserve Enter Agreement." *Thomasville (GA) Times-Enterprise*, December 24, 2019. www.timesenterprise.com.

## Chapter 4. River Otters

Behnke, Patricia C. *Old Timers Remember: Ichetucknee Springs*. Florida Department of Environmental Protection, June 2003.

Center for Biological Diversity. "Saving the Ichetucknee Silt Snail." Accessed February 9, 2024. www.biologicaldiversity.org/species/invertebrates /Ichetucknee siltsnail/index.html.

Cornell Lab of Ornithology. "Limpkin." All About Birds. Accessed February 14, 2024. www.allaboutbirds.org/guide/Limpkin/overview#.

Florida Fish and Wildlife Conservation Commission. "Florida's Apple Snails." Accessed February 14, 2024. https://myfwc.com/research/freshwater/species -assessments/mollusks/apple-snails/.

Grahame, Kenneth. *The Wind in the Willows*. Ware, England: Wordsworth Editions, 1998, 65. First published 1908.

Ichetucknee Alliance. "An Ichetucknee Historical Timeline." Ichetucknee Beloved Blue River Project. 2023. https://belovedblueriver.org/the-human-role/history -and-long-term-trends/historical-timeline/.

Keller, Anett. "Why We Should Protect Karst Landscapes." Heinrich Böll Stiftung Southeast Asia, March 3, 2021. https://th.boell.org.

Merritt, Lu, and Angeline Meeks. "Exploring the Ichetucknee Springshed." Ichetucknee Beloved Blue River Project. Accessed February 9, 2024. https:// storymaps.arcgis.com/stories/f38b0f84f34a42e0941416fd873133b8.

Merritt, Lucinda Faulkner. "Interview with Jim Stevenson." Ichetucknee Alliance, June 17, 2019. http://ichetuckneealliance.org/wp-content/uploads/2020/04 /Interview-with-Jim-Stevenson-FINAL.pdf. (The interview is no longer available online but can be requested from the Ichetucknee Alliance.)

Milanich, Jerald T. *Florida Indians and the Invasion from Europe*. Gainesville: University Press of Florida, 1993.

Mirocha, Anna. "Save the Weirdos—A Snail's Tale: The Itsy-bitsy, Teenie-weenie Ichetucknee Siltsnail." Center for Biological Diversity. December 23, 2016. https://medium.com/center-for-biological-diversity/who-isnt-a-little-weird -7b3332a06dc3.

The Nature Conservancy. "River Otters." Accessed February 8, 2024. www.nature .org/en-us/about-us/where-we-work/united-states/indiana/stories-in -indiana/welcome-back-river-otters/.

Rawlings, Marjorie Kinnan. *Cross Creek*. New York: Scribner, 1942, 244.

Ringle, Ken. "North Florida Springs." *National Geographic*, March 1999, 40–59.

Roberts, Nathan M., Matthew J. Lovallo, and Shawn M. Crimmins. "River Otter Status, Management, and Distribution in the United States: Evidence of

Large-Scale Population Increase and Range Expansion." *Journal of Fish and Wildlife Management* 11, no. 1 (2020): 279–86. https://doi.org/10.3996/102018 -JFWM-093.

State of Florida Department of Environmental Protection Division of Recreation and Parks. *Ichetucknee Springs State Park Approved Unit Management Plan.* October 17, 2000. https://floridadep.gov/sites/default/files/2000%20 Ichetucknee%20Springs%20State%20Park.pdf.

Traynor, Jeff. "Conservation Success: River Otter Populations Continue to Thrive in America." Furbearer Conservation. June 21, 2020. https://furbearer conservation.com/blog/2020/6/19/river-otter-populations-continue-to-thrive -in-america.

Weary, David J., and Daniel H. Doctor. *Karst in the United States: A Digital Map Compilation and Database*: US Geological Survey Open-File Report 2014-1156. Reston, VA: US Geological Survey, 2014. http://dx.doi.org/10.3133/ofr20141156.

Weisman, Brent Richards. *Excavations on the Franciscan Frontier: Archeology at the Fig Springs Mission.* Gainesville: University Press of Florida / Florida Museum of Natural History, 1992.

Yang, Qian-Qian, Su-Wen Liu, Chao He, and Xiao-Ping Yu. "Distribution and the Origin of Invasive Apple Snails, *Pomacea canaliculata* and *P. maculata* (Gastropoda: Ampullariidae) in China." *Scientific Reports* 8, no. 1185 (2018). www.nature.com/articles/s41598-017-19000-7.

## Chapter 5. Alligators

Bell, Vereen. *Swamp Water.* Athens: University of Georgia Press, 1981.

Berryhill, Don. "Episode 21: Okefenokeology: Oscar the Alligator." YouTube video, 25:50. Posted by Okefenokee Swamp Park on May 13, 2020. www.youtube.com/watch?v=QBzxbZrvSzo.

Georgia Conservancy. "Mining Threatens the Okefenokee Swamp." Accessed February 15, 2024. www.georgiaconservancy.org/okefenokee/mining.

Georgia River Network. "The Okefenokee Is Endangered." Accessed February 8, 2024. https://garivers.org/protectokefenokee/.

Harper, Francis, and Delma E. Presley. *Okefinokee Album.* Athens: Brown Thrasher Books, University of Georgia Press, 1981.

Hoog, Mark. "The Effect of Genetic Relatedness on Mate Selection and Spatial Distribution in the American Alligator, *Alligator mississippiensis*." Master's thesis, Georgia Southern University, 2023. https://digitalcommons.georgia southern.edu/etd/2619.

Jackson, Gordon. "Okefenokee's Oscar Expected to Be a Big Attraction Even in Death." *Florida Times-Union*, July 19, 2009. www.jacksonville.com/story /news/2009/07/19/okefenokees-oscar-expected-to-be-a-big-attraction-even -in-death/15979106007/.

Landers, Mary. "Okefenokee on Track to World Heritage Site Status." *Georgia Recorder*, September 27, 2023. https://georgiarecorder.com/2023/09/27/okefenokee-on-track-to-world-heritage-site-status/.

Lenz, Richard J. "Flora of the Okefenokee Swamp." Sherpa Guides. The Natural Georgia Series: Okefenokee Swamp. Accessed February 9, 2024. www.sherpaguides.com/georgia/okefenokee_swamp/wildnotes/index.html.

Nelson, Megan Kate. "The Okefenokee Swamp (Georgia/Florida)." In *Southern United States: An Environmental History*, by Donald E. Davis, 221–34. Santa Barbara, CA: ABC CLIO, 2006.

Ray, Janisse. "I Have Seen the Warrior: How Not to Die." *Wild Spectacle: Seeking Wonders in a World Beyond Humans*. San Antonio: Trinity University Press, 2021.

Savannah River Ecology Laboratory. "How to Be Safe around Alligators." Accessed February 9, 2024. https://srelherp.uga.edu/alligators/alligator-safety.htm.

Simmons, Kelly. "UGA Partners with Okefenokee Swamp to Conserve and Protect Native Alligators." August 13, 2020. https://outreach.uga.edu/uga-partners-with-okefenokee-swamp-to-conserve-and-protect-native-alligators/.

University of Georgia Marine Grant Extension and Georgia Sea Grant. "Alligator Research: What We Have Learned." Accessed February 8, 2024. https://okeswamp.org/alligator-research/what-we-have-learned/.

## Chapter 6. Frogs and Toads

Camargo, Suzana. "Study Shows Dire Outlook for Amphibians: 40 Percent Threatened with Extinction." *Mongabay*. November 20, 2023. https://news.mongabay.com/2023/11/study-shows-dire-outlook-for-amphibians-40-threatened-with-extinction/.

Dillard, Annie. *Pilgrim at Tinker Creek*. New York: Harper Perennial Modern Classics, 2009, 7.

Dorcas, Mike, and Whit Gibbons. *Frogs and Toads of the Southeast*. Wormsloe Foundation Nature Books. Athens: University of Georgia Press, 2008.

Grandoni, Dino. "Climate Change Is Driving Many Amphibians toward Extinction." *Washington Post*, October 4, 2023. www.washingtonpost.com/climate-environment/2023/10/04/frog-climate-amphibians-extinction.

Ivey, Jennie. "What's That Shining in the Grass?" *Cookeville Herald Citizen*, October 27, 2019. www.jennieivey.com/newspaper-columns/whats-that-shining-in-the-grass/.

Lovett, Gary M. "When Do Peepers Peep? Climate and the Date of First Calling in the Spring Peeper (*Pseudacris crucifer*) in Southeastern New York State." *Northeastern Naturalist* 20, no. 2 (2013): 333–40. www.eaglehill.us/NENAonline/articles/NENA-20-2/18-Lovett.shtml.

Savannah River Ecology Laboratory, University of Georgia. "Frogs and Toads of

South Carolina and Georgia." Accessed August 10, 2024. https://srelherp.uga
.edu/frogs-and-toads/.

US Geological Survey. "USGS Frog Quiz." Patuxent Wildlife Research Center.
January 30, 2015. www.pwrc.usgs.gov/frogquiz/index.cfm?fuseaction=main
.lookup.

## Chapter 7. Eastern Screech-Owls

Einhorn, Catrin. "The Mystery of the Vanishing Kestrels: What's Happening to
This Flash Falcon?" *New York Times*, June 5, 2023. www.nytimes.com/2023
/06/05/climate/american-kestrel.html.

Jacobus, Luke M., Craig R. Macadam, and Michel Sartori. "Mayflies
(Ephemeroptera) and Their Contributions to Ecosystem Services." *Insects* 10,
no. 6 (June 2019): 170. www.ncbi.nlm.nih.gov/pmc/articles/PMC6628430/.

Kooser, Ted. "Screech Owl." *Delights and Shadows*. Port Townsend, WA: Copper
Canyon, 2004, 73.

Morton, Oren F. *A History of Highland County, Virginia*. Monterey, VA: published
by the author, 1911. https://archive.org/details/historyofhighlanoomort/page
/n5/mode/2up.

Podger, Pamela J. "In a Corner of Virginia's 'Switzerland,' a Division over a
Planned Wind Farm." *New York Times*, February 13, 2007. www.nytimes
.com/2007/02/13/us/13wind.html.

Reum, Patti, and John Spahr. "American Kestrel Project in Virginia." Brandywine
Zoo, Wilmington, DE, 2014. Accessed August 10, 2024. https://brandywinezoo
.org/wp-content/uploads/2017/05/REUM-AMKE-Sym-Presentation-Final
-Version.pdf.

Spahr, John. "The Eastern Screech-Owl (*Megascops asio*) in Highland County,
Virginia: A Study of Its Prevalence and Distribution." *Raven: Journal of the
Virginia Society of Ornithology* 86, no. 2 (2015): 3–8. https://corn-tortoise-83wr
.squarespace.com/s/862-2015.pdf.

USDA National Agricultural Statistics Service. "Highland County Virginia,
Census of Agriculture." 2017. www.nass.usda.gov/Publications/AgCensus
/2017/ Online_Resources/County_Profiles/Virginia/cp51091.pdf.

## Chapter 8. Dismalites

Brandes, Heidi. "See Glowworms Give a Dazzling Show at This National
Landmark." *National Geographic*, August 30, 2021. www.nationalgeographic
.com/travel/article/see-dismalites-light-up-dismal-canyon-natural-area.

Coder, Kim D. "Fox-fire Makes Forests Glow." CAES Newswire, College of
Agricultural and Environmental Science, UGA Cooperative Extension,
February 18, 2010. https://newswire.caes.uga.edu/story/3715/fox-fire-glow
.html.

Faust, Lynn Frierson. *Fireflies, Glow-worms, and Lightning Bugs: Identification and Natural History of the Fireflies of the Eastern and Central United States and Canada.* Wormsloe Foundation Nature Books. Athens: University of Georgia Press, 2017.

Fulton, B. B. "Lochetic Luminous Dipterous Larvae." *Journal of the Elisha Mitchell Science Society* 55, no. 2 (December 1939): 289–93. https://dc.lib.unc.edu/cdm/ref/collection/jncas/id/1757.

Hudson, Steve. "By the Light of a Glow Worm." *Alpharetta and Roswell News,* July 11, 2021. appenmedia.com.

Kanuckel, Amber. "Glowworms: The Crazy World of Dismalites." *Farmer's Almanac* 31 (January 2024). www.farmersalmanac.com/glow-worms.

Lindenberg, Andrea. "Hackleburg Was Never the Same after the Tornado. A Decade Later Residents Continue to Lean on One Another." CBS42.com, April 27, 2021. www.cbs42.com/special-reports/hackleburg-residents-leaned-on-one-another-to-get-through-tornado-10-years-ago/.

Nakashima, Karen. *Glow: Living Lights.* Teacher's Guide, San Diego Natural History Museum. Accessed August 10, 2024. www.sdnhm.org.

National Geographic Society. "Light Pollution." National Geographic Education. October 19, 2023. https://education.nationalgeographic.org/resource/light-pollution/.

———. "The Bioluminescence Webpage." National Geographic Education. October 19, 2023. https://education.nationalgeographic.org/resource/bioluminescence/.

Renfranz, Amy. "Dear Naturalist: The Blue Glow of the Orfelia fultoni." *Watauga (NC) Democrat,* May 18, 2017. https://wataugademocrat.com.

Richardson, James. "Nothing Dismal about Dismals Canyon." *Alabama Heritage,* September 29, 2016. www.alabamaheritage.com/from-the-vault/-nothing-dismal-about-dismals-canyon.

Sivinski, John. "Arthropods Attracted to Luminous Fungi." *Psyche: A Journal of Entomology* 88, nos. 3–4 (January 1981). https://doi.org/10.1155/1981/79890.

———. "Prey Attraction by Luminous Larvae of the Fungus Gnat *Ofelia fultoni.*" *Ecological Entomology* 7, no. 4 (November 1982): 443–46. https://doi.org/10.1111/j.1365-2311.1982.tb00686.x.

Smithsonian Institution Archives. "Bentley Ball Fulton (1889–1960)." Accessed August 10, 2024. https://siarchives.si.edu/collections/siris_arc_382990.

Thornton, William. "Tornadoes, Moment-by-Moment: How It Happened." AL.com, April 27, 2021. www.al.com/news/2021/04/april-27–2011-tornadoes-moment-by-moment-how-it-happened.html.

Viviani, Vadim R., Jaqueline R. Silva, Danilo T. Amaral, Vanessa R. Bevilaqua, Fabio C. Abdalla, Bruce R. Branchini, and Carl H. Johnson. "Author Correction:

A New Brilliantly Blue-Emitting Luciferin-Luciferase System from *Orfelia
fultoni* and Keroplatinae (Diptera)." *Scientific Reports* 13, no. 4649 (March
2023). https://rdcu.be/dykBO.

Whitman, Walt. "Miracles." In *Poems by Walt Whitman*, edited by William
Michael Rosetti, 212. London: Chatto & Windus, 1901.

## Chapter 9. Fireflies

Associated Press. "Elf Lobby Blocks Iceland Road Project." *The Guardian*,
December 22, 2013. www.theguardian.com/world/2013/dec/22/elf-lobby
-iceland-road-project.

Chávez, Karen. "Discovery of Synchronous Fireflies at Grandfather Mountain
Could Help Ease Smokies Crowds." *Asheville (NC) Citizen Times*, September 3,
2019. www.citizen-times.com/story/life/2019/09/03/synchronous-fireflies
-have-been-discovered-grandfather-mountain-nc/2165336001/.

———. "Where in Western North Carolina Can You See the Blue Ghost
Fireflies?" *Asheville (NC) Citizen Times*, May 11, 2018. www.citizen-times.com
/story/news/local/2018/05/11/spot-blue-ghost-fireflies-around-asheville
-pisgah-dupont-forests/600064002/#.

Faust, Lynn F., and Timothy G. Forrest. "Bringing Light to the Lives of the
Shadow Ghosts, *Phausus inaccensa* (*Coleoptera:Lampyridae*)" *American
Entomologist* 63, no. 3 (Fall 2017): 177–89. https://doi.org/10.1093/ae/tmx027.

Faust, Lynn Frierson. *Fireflies, Glow-worms, and Lightning Bugs: Identification
and Natural History of the Fireflies of the Eastern and Central United States and
Canada*. Wormsloe Foundation Nature Books. Athens: University of Georgia
Press, 2017.

Frick-Ruppert, Jennifer E., and Joshua J. Rosen. "Morphology and Behavior
of *Phausis reticulata* (Blue Ghost Firefly)." *Journal of the North Carolina
Academy of Science 124, no. 4* (2008): 139–47. https://dc.lib.unc.edu/cgi-bin
/showfile.exe?CISOROOT=/jncas&CISOPTR=3883.

Galton, Tal. "Magical Fireflies of the Blue Ridge Mountains." *Snakeroot Ecotours*
(blog). June 5, 2021. www.snakerootecotours.com/post/magical-fireflies-of
-the-southern-appalachians.

Grandfather Mountain Stewardship Foundation. "Firefly Species on
Grandfather." Accessed February 9, 2024. https://grandfather.com/firefly
-species/.

Hubbard, Harlan. *Payne Hollow Journal*. Lexington: University Press of Kentucky,
1996, 42.

Sorenson, Clyde. "Hunting Carolina Ghosts." *Our State*, March 27, 2023. www.our
state.com/hunting-carolina-ghosts/.

Valich, Lindsey. "Firefly Researchers Mapping 'World's Second-Most Interesting

Genome.'" University of Rochester News Center, September 10, 2017. www
.rochester.edu/newscenter/firefly-researchers-mapping-worlds-second-most
-interesting-genome-269372/.

## Chapter 10. Purple Martins

Audubon, John James. *Birds of America*. New York: Welcome Rain, 2001.

Brown, Charles R. "Sleeping Behavior of Purple Martins." *Condor* 82, no. 2 (1980):
170–75. https://doi.org/10.2307/1367472.

Coates, Peter. *American Perceptions of Immigrant and Invasive Species: Strangers
on the Land*. Berkeley: University of California Press, 2006.

Cole, Adam, and Maggie Starbard, prod. "The Mystery of the Missing Martins |
Field Trip! | Skunk Bear." *Skunk Bear*, NPR. YouTube video, 08:35. Posted by
NPR's Skunk Bear on December 4, 2014. https://www.youtube.com/watch
?v=SoHP056_Gfc.

Connor, Jack. "Purple Martins, Ecological Mismatches, and Climate Change."
*Living Bird*, Spring 2015. www.allaboutbirds.org/news/purple-martins
-ecological-mismatches-and-climate-change/.

Dickinson, Phil, and Ron Morris. "Martin Migration Spectacular to Behold
in N.C., Locals Get Involved as Innkeepers." *Winston-Salem (NC) Journal*,
April 16, 2021. https://journalnow.com.

Eggers, Caroline. "Nashville Symphony Cuts Trees in Anticipation of Purple
Martin Invasion." WPLN.org, May 17, 2022. https://wpln.org/post/nashville
-symphony-cuts-trees-in-anticipation-of-purple-martin-invasion/.

Fonner, Anna M. "Gleanings from the Life of J. Warren Jacobs, Scientist." *Western
Pennsylvania Historical Magazine* 50, no. 3 (July 1967): 214–20. https://journals
.psu.edu/wph/article/view/2875/2707.

Fugate, Lauren, and John MacNeill Miller. "Shakespeare's Starlings: Literary
History and the Fictions of Invasiveness." *Environmental Humanities* 13, no. 2
(November 2021): 301–22. https://doi.org/10.1215/22011919–9320167.

Gershon, Livia. "The Disappearing Culture of Purple Martin Landlords." JSTOR
Daily. January 21, 2021. https://daily.jstor.org/the-disappearing-culture-of
-purple-martin-landlords/.

Grossman, Daniel. "This Tiny Brazilian Island Could Hold the Key to the Purple
Martin's Future." *Audubon Magazine*, Fall 2022. https://www.audubon.org
/magazine/fall-2022/this-tiny-brazilian-island-could-hold-key-purple.

Helms, Jackson A., Aaron P. Godfrey, Tanya Ames, and Eli S. Bridge. "Are Invasive
Fire Ants Kept in Check by Native Aerial Insectivores?" *Biology Letters* 12, no. 5
(May 2016). https://royalsocietypublishing.org/doi/epdf/10.1098/rsbl.2016
.0059.

Jervis, Lori L., Paul Spicer, William C. Foster, Jeffrey Kelly, and Eli Bridge.

"Resisting Extinction: Purple Martins, Death, and the Future." *Conservation & Society* 17, no. 3 (2019): 227–35. www.jstor.org/stable/26677959.

Knittle, C. E., J. L. Guarino, P. C. Nelson, R. W. Dehaven, and D. J. Twedt. "Baiting Blackbird and Starling Congregating Areas in Kentucky and Tennessee." *Proceedings of the 9th Vertebrate Pest Conference* (March 1980): 31–37. https://digitalcommons.unl.edu/vpc9/20.

Magris, Claudio. *Danube: A Sentimental Journey from the Source to the Black Sea.* Toronto: Harvill, 1990.

Mott, Donald F. "Dispersing Blackbirds and Starlings from Objectionable Roost Sites." *Proceedings of the 9th Vertebrate Pest Conference* (March 1980): 38–42. https://digitalcommons.unl.edu/cgi/viewcontent.cgi?article=1027&context=vpc9.

Murray, Brian T. "Dead Birds in Franklin Township Were Killed on Purpose." *Star-Ledger* (Newark, NJ), January 26, 2009. www.nj.com/news/2009/01/dead_birds_littering_franklin.html.

NC Purple Martin Society. "Purple Martin Info and Management." Accessed February 10, 2024. www.ncpurplemartin.org/about-purple-martins#.

Peterson, Bo. "Pooping Purple Birds Pit Neighbor vs. Neighbor in Hanahan Neighborhood." *Charleston Post and Courier*, September 14, 2020. www.postandcourier.com.

Purple Martin Conservation Association. Accessed February 10, 2024. www.purplemartin.org.

Renkl, Margaret. "A Flock of Beautiful Birds in a City Is a Miracle, a Disaster and a Conundrum." *New York Times*, April 11, 2022. www.nytimes.com/2022/04/11/opinion/nashville-symphony-bird-migration.html.

———. "The Greatest Show in Nashville Is Happening in the Sky." *New York Times*, August 7, 2023. www.nytimes.com/2023/08/07/opinion/purple-martins-birds-nashville.html.

———. "A 150,000-Bird Orchestra in the Sky." *New York Times*, September 7, 2020. www.nytimes.com/2020/09/07/opinion/a-150000-bird-orchestra-in-the-sky.html.

Richmond, Stanley M. "The Attraction of Purple Martins to an Urban Location in Western Oregon." *Condor* 55, no. 5 (1953): 225–49. https://doi.org/10.2307/1365035.

Ripp, Rachel. "Filmmaker Creates Conservation Documentary about Lake Murray Purple Martins." WLTX.com, April 22, 2022.

Stutchbury, Bridget J. M. "Coloniality and Breeding Biology of Purple Martins (*Progne subis hesperia*) in Saguaro Cacti." *Condor* 93, no. 3 (1991): 666–75. https://doi.org/10.2307/1368198.

———. "Tracking Purple Martin Migration to Brazil, and Back!" *Purple Martin*

*Update: A Quarterly Journal* 18, no. 2 (n.d.): 2–6. www.purplemartin.org
/uploads/media/18-2-geolocators-345.pdf.

Wilson, Alexander, C. L. Bonaparte, R. Jameson, G. Ord, and W. M. Hetherington.
*American Ornithology; or the Natural History of the Birds of the United States.*
Philadelphia: Bradford and Inskeep, 1831. www.biodiversitylibrary.org
/bibliography/97204.

## Chapter 11. Moths and Butterflies

Bargar, Timothy A., Michelle L. Hladik, and Jaret C. Daniels. "Uptake
and Toxicity of Clothianidin to Monarch Butterflies from Milkweed
Consumption." *PeerJ* 10, no. 8 (March 2020). www.ncbi.nlm.nih.gov/pmc
/articles/PMC7069410/.

Bargman, Joe. "Flying Colors: A Virginia Painter Elevates the Profile of the Lowly
Moth." *Garden and Gun*, August/September 2017. https://gardenandgun.com
/articles/flying-colors-moth-paintings/.

Buck, Brad. "UF Project Hopes to 'Certify' Appropriate Plants as Wildlife
Friendly." University of Florida, June 17, 2021. https://blogs.ifas.ufl.edu
/news/2021/06/17/uf-project-hopes-to-certify-appropriate-plants-as-wildlife
-friendly/.

Cody, John. "John Cody Gallery: Audubon of Moths." Accessed February 10, 2024.
www.johncodygallery.net/about-artist/.

Daniels, Jaret. *Native Plant Gardening for Birds, Bees and Butterflies: Southeast.*
Cambridge, MN: Adventure Publications, 2021.

———. *Vibrant Butterflies: Our Favorite Visitors to Flowers and Gardens.*
Cambridge, MN: Adventure Publications, 2018.

Daniels, Jaret, Chase Kimmel, Simon McClung, Samm Epstein, Jonathan Bremer,
and Kristin Rossetti. "Better Understanding the Potential Importance of
Florida Roadside Breeding Habitat for the Monarch." *Insects* 9, no. 4 (October
2018): 137. www.ncbi.nlm.nih.gov/pmc/articles/PMC6315611/.

First Magnitude Brewing Company. "Saving Species Never Tasted So Good."
YouTube video, 07:57. Posted by Florida Museum on October 31, 2018. https://
www.youtube.com/watch?v=DCR-OdfOowo.

Grace, Kristen. "A Kaleidoscope of Monarchs: Marveling at One of Nature's
Greatest Journeys." Florida Museum, October 29, 2019. www.floridamuseum
.ufl.edu/science/marveling-at-monarchs/.

Higgins, Adrian. "The Underrated Beauty of Moths." *Washington Post*, August 23,
2016. www.washingtonpost.com.

Hill, Geena M., Akito Y. Kawahara, Jaret C. Daniels, Craig C. Bateman, and
Brett R. Scheffers. "Climate Change Effects on Animal Ecology: Butterflies and
Moths as a Case Study." *Biological Reviews* 96 no. 5 (October 2021): 2113–26.
https://onlinelibrary.wiley.com/doi/10.1111/brv.12746.

Himmelman, John. *Discovering Moths: Nighttime Jewels in Your Own Backyard.* Mechanicsburg, PA: Stackpole Books, 2023.

Kawahara, Akito Y., Lawrence E. Reeves, Jesse R. Barber, and Scott H. Black. "Eight Simple Actions That Individuals Can Take to Save Insects from Global Declines." *PNAS* 118, no. 2 (January 11, 2021). https://doi.org/10.1073/pnas.2002547117.

Kingsolver, Barbara. *Prodigal Summer.* New York: Harper Perennial, 2001.

Leckie, Seabrooke, and David Beadle. *Peterson Field Guide to Moths of Southeastern North America.* Boston: Houghton Mifflin Harcourt, 2018.

Nanz, Steve, ed. "Digital Guide to Moth Identification." North American Moth Photographers Group at the Mississippi Entomological Museum at Mississippi State University. Accessed February 10, 2024. http://mothphotographersgroup.msstate.edu.

Pinson, Jerald. "Florida Volunteers See Record Numbers of Endangered Schaus' Swallowtail Butterfly." Florida Museum, September 16, 2021. www.floridamuseum.ufl.edu/science/record-numbers-of-floridas-endangered-schaus-swallowtail-butterfly/.

———. "What Is It Good For? Absolutely One Thing. Luna Moths Use Their Tails Solely for Bat Evasion." Florida Museum, April 10, 2023. www.floridamuseum.ufl.edu/science/what-is-it-good-for-absolutely-one-thing-luna-moths-use-their-tails-solely-for-bat-evasion/.

Smithsonian Institution. "Moths." BugInfo: Information Sheet Number 169. Accessed February 12, 2024. www.si.edu/spotlight/buginfo/moths.

Thompson, Joanna. "To Kill or Not to Kill: Butterflying during the 'Insect Apocalypse.'" *The Highlight by Vox*, May 24, 2022. www.vox.com/the-highlight/23055318/butterfly-collection-insect-climate-change.

Sylvestri, Brittany. "Brewing Beer for Butterflies: UF Alumni, Professor Partner to Save Declining Species." University of Florida News, August 4, 2023. https://news.ufl.edu/2023/08/beer-and-butterflies/.

Wagner, David L., Richard Fox, Danielle M. Salcido, and Lee A. Dyer. "A Window to the World of Global Insect Declines: Moth Biodiversity Trends Are Complex and Heterogeneous." *PNAS* 118, no. 2 (January 11, 2021). https://doi.org/10.1073/pnas.2002549117.

## Chapter 12. Forest Bathing

Beresford-Kroeger, Diana. *To Speak for the Trees.* Toronto: Random House Canada, 2019.

Brincka, Matt. "The Ghost Plant: A Closer Look at the Spookiest Plant in the Forest." *New York State Parks and Historic Sites Blog*, August 28, 2018. https://nystateparks.blog/2018/08/28/.

Global Wellness Institute. "Forest Bathing Spotlight." Accessed August 10, 2024.

https://globalwellnessinstitute.org/wellnessevidence/forest-bathing/forest
-bathing-spotlight/.

Jackson-Buckley, Bridgitte. "What Is Plant Spirit Medicine?" Gaia, April 26, 2017.
www.gaia.com/article/what-plant-spirit-medicine.

Li Quing, M. Kobayashi, H. Inagaki, Y. Hirata, Y. J. Li, K. Hirata, T. Shimizu, et al.
"A Day Trip to a Forest Park Increases Human Natural Killer Activity and the
Expression of Anti-Cancer Proteins in Male Subjects. *Journal of Biological
Regulators and Homeostatic Agents* 24, no. 2 (April–June 2010): 157–65.
https://pubmed.ncbi.nlm.nih.gov/20487629/.

Livni, Ephrat. "Japanese 'Forest Medicine' Is the Science of Using Nature to
Heal Yourself—Wherever You Are." Quartz, February 21, 2018. https://qz
.com/1208959/japanese-forest-medicine-is-the-art-of-using-nature-to-heal
-yourself-wherever-you-are.

Marie, Simone. "Sand Tray Therapy: Evidence, Benefits, and More." Healthline,
September 7, 2022. www.healthline.com/health/mental-health/sand-tray
-therapy.

Park, Bum Jin, Yuko Tsunetsugu, Tamami Kasetani, Takahide Kagawa, and
Yoshifumi Miyazaki. "The Physiological Effects of Shinrin-yoku (Taking In the
Forest Atmosphere or Forest Bathing): Evidence from Field Experiments in 24
Forests across Japan." *Environmental Health and Preventive Medicine* 15, no. 1
(January 2010): 18–26. www.ncbi.nlm.nih.gov/pmc/articles/PMC2793346/.

Sorg, Lisa. "Over the Past Decade Wake County Lost 11,120 Acres of Trees—
Equivalent to 2,700 Walmart Stores." NC Newsline, September 7, 2023.
https://ncnewsline.com/2023/09/07/.

Weil, Simone. *Gravity and Grace.* Lincoln: University of Nebraska Press, 1997.

## Chapter 13. Wood Storks and Roseate Spoonbills

American Bird Conservancy. "ABC's Bird Library: Roseate Spoonbill." Accessed
February 12, 2024. https://abcbirds.org/bird/roseate-spoonbill/.

Beaufort Life. "Last Cabin Standing at Hunting Island Coming Down Forever."
Eat, Stay, Play, Beaufort. Accessed February 12, 2024. www.eatstayplaybeaufort
.com/last-cabin-standing-at-hunting-island-coming-down-forever/.

Bleser, Carol, ed. *Secret and Sacred: The Diaries of James Henry Hammond,
a Southern Slaveholder.* New York: Oxford University Press, 1989.

Bodine, Renee. "Saving the Wood Stork: Engineering a Revival across the
Southeast." US Fish and Wildlife Service, June 24, 2022. https://www.fws.gov
/story/2022-06/saving-wood-stork.

Brown, Rosellen. "Monster of All He Surveyed." *New York Times,* January 29,
1989, section 7, 22. www.nytimes.com/1989/01/29/books/monster-of-all-he
-surveyed.html.

Center for Biological Diversity. "Status and Recovery of the Wood Stork (*Mycteria americana*)." January 2012. www.biologicaldiversity.org/species/birds/wood _stork/status_study.html.

Coulter, Malcolm C., William D. McCort, and A. Lawrence Bryan. "Creation of Artificial Foraging Habitat for Wood Storks." *Colonial Waterbirds* 10, no. 2 (1987): 203–10. https://doi.org/10.2307/1521259.

Fallaw, W. C., David S. Snipes, and Van Price. "Wandering with William Bartram: The Section at Silver Bluff, South Carolina." *Earth Sciences History* 13, no. 1 (1994): 52–57. www.jstor.org/stable/24137325.

Faust, Drew Gilpin. *James Henry Hammond and the Old South: A Design for Mastery*. Baton Rouge: LSU Press, 1982.

Hamel, Paul B. "The Wood Stork in South Carolina, A Review." *The Chat*, Spring 1977, 24–27. www.carolinabirdclub.org/chat/issues/1977/v41n2wost.pdf.

Hooles, Ken. "Birdwatch: Bird from Texas and Florida Visits Ontario Town." *Pembroke Observer* (Toronto), October 3, 2023. www.pembrokeobserver .com/opinion/columnists/birdwatch-rare-bird-from-texas-and-florida-visits -ontario-town.

Lanham, Drew. *The Home Place: Memoirs of a Colored Man's Love Affair with Nature*. Minneapolis: Milkweed Editions, 2016.

McMenamin, Kathleen. "The Roseate Spoonbill." Hiltonhead.com. Accessed February 16, 2024. www.hiltonhead.com.

Miles, Susannah Smith. "Follow the Colorful Life of the Roseate Spoonbill." *Charleston Magazine*, November 2020. https://charlestonmag.com/features /follow_the_colorful_life_of_the_roseate_spoonbill.

Murphy, Thomas M., and Christine E. Hand. "Wood Stork (*Mycteria americana*)." *South Carolina State Wildlife Action Plan 2015*, 2015 Supplemental Volume. SC Department of Natural Resources. www.dnr.sc.gov/swap/supplemental/birds /woodstork2015.pdf.

Murphy, Thomas M., and John W. Coker. "A Twenty-Six Year History of Wood Stork Nesting in South Carolina." *Waterbirds* 31, Special Publication 1 (2008): 3–7. www.dnr.sc.gov/wildlife/species/wadingbirds/docs/WoodStorkTNesting History.pdf.

National Park Service. "Wood Stork: Species Profile." Everglades National Park, Florida, April 8, 2021. www.nps.gov/ever/learn/nature/woodstork.htm.

Pardue, Doug. "Deadly Legacy: Savannah River Site Near Aiken One of the Most Contaminated Places on Earth." *Post and Courier* (South Carolina), May 21, 2017. www.postandcourier.com.

Rackley, Kristina. "Silver Bluff Audubon Is a Sanctuary in More Ways Than One." *Aiken Standard* (Aiken County, SC), July 28, 2019. https://sc.audubon.org /news/silver-bluff-audubon-sanctuary-more-ways-one.

Richardson, Natalie Rose. "Look Close or You'll Miss It." *Emergence Magazine*, September 14, 2023. https://emergencemagazine.org/essay/look-closely-or-youll-miss-it/.

*SCIWAY News*. "Cooling Tower K Reactor—Savannah River Site South Carolina." May 2009. Accessed February 12, 2024. www.sciway.net/srs-savannah-river-site/.

South Carolina Department of Archives and History. "Silver Bluff, Aiken County (Address Restricted)." National Register Properties in South Carolina. Accessed February 12, 2024. www.nationalregister.sc.gov/aiken/S10817702008/index.htm.

South Carolina Department of Natural Resources. "SCDNR Notes Record Number of Wood Stork Nests in 2022." SCDNR News, December 5, 2022. www.dnr.sc.gov/news/2022/dec/dec5-woodstork.php.

Sutton-Jackson, Vicky L. "SREL, DOE Bring Wood Stork Back from Brink of Extinction." Savannah River Ecology Laboratory, University of Georgia. Accessed February 12, 2024. https://srel.uga.edu/srel-doe-bring-wood-stork-back-from-brink-of-extinction/.

Ulrich, Carey. "Hunting Island's Erosion Problem: Can It Be Saved or Is It Doomed to Disappear?" April 27, 2021 (no longer available).

## Chapter 14. Bugling Elk

Brill, David. "A Graceful Giant Makes a Successful Return." *Smokies Life Magazine*, Fall 2021, 30–43.

Carnes, Phoebe. "Echoes in the Mountains: On Bulls and Close Encounters." *Smokies LIVE Blog*, Smokies Life, February 21, 2022. https://smokieslife.org/2022/02/21/echoes-in-the-mountains-on-bulls-and-close-encounters/.

Clapsaddle, Annette Saunooke. *Even as We Breathe*. Lexington: University Press of Kentucky, 2020.

Igelman, Jack. "Elk and Other Large Species in NC Mountains Constrained by Changing Habitats, Human Activity." Carolina Public Press, May 25, 2022. https://carolinapublicpress.org/54022/elk-and-other-large-species-in-nc-mountains-constrained-by-changing-habitats-human-activity/.

Juniper, Robert. "Commentary: Menagerie." Cherokee One Feather, February 21, 2021. https://theonefeather.com/2022/02/21/commentary-menagerie/.

Mendez, Brittany. "The Story of Elk in the Great Smoky Mountains." Great Smoky Mountains National Park blog, September 1, 2021. https://smokymountains.com/park/blog/story-elk-great-smoky-mountains/.

Morgan, Robert. "Writing the Mountains." *Cornell Plantations Magazine*. Vol. 52. Ithaca, NY: Cornell University Press, 1997.

Murrow, Jennifer L., Joseph D. Clark, and E. Kim Delozier. "Demographics of an Experimentally Released Population of Elk in Great Smoky Mountains

National Park." *Journal of Wildlife Management* 73, no. 8 (2009): 1261–68. www.jstor.org/stable/20616792.

Quinlan, Braiden A., Jacalyn P. Rosenberger, David M. Kalb, Heather N. Abernathy, Emily D. Thorne, W. Mark Ford, and Michael J. Cherry. "Drivers of Habitat Quality for a Reintroduced Elk Herd." *Scientific Reports* 12, no. 20960 (2022). www.nature.com/articles/s41598-022-25058-9.

Searcy, Aaron. "Word from the Smokies: Wildlife Biologist Helps Elk Return to Appalachia." *Asheville Citizen Times*, September 11, 2021. www.citizen-times.com/story/news/local/2021/09/11/word-smokies-wildlife-biologist-helps-elk-return-appalachia/5777054001/.

Silver, Timothy. *Mount Mitchell and the Black Mountains: An Environmental History of the Highest Peaks in Eastern America.* Chapel Hill: University of North Carolina Press, 2003, 193–94.

Simmons, Morgan. "Smokies Park Struggling to Clamp Down on Hog Infestation." *Knoxville (TN) News*, January 23, 2010. www.knoxnews.com.

Taylor, Mark. "Virginia's Expanding Elk Herd a Growing Draw for Buchanan County." *Cardinal News* (Virginia), February 17, 2022. https://cardinalnews.org/2022/02/17/virginias-expanding-elk-herd-a-growing-draw-for-buchanan-county/.

Van Doren, Mark, ed. *Travels of William Bartram.* New York: Dover Publications, 1928.

Witmer, Gary. "Reintroduction of Elk in the United States." *Journal of the Pennsylvania Academy of Science* 64, no. 3 (1990): 131–35. www.jstor.org/stable/44148971.

Yarkovich, Joseph, Joseph D. Clark, and Jennifer L. Murrow. "Effects of Black Bear Relocation on Elk Calf Recruitment at Great Smoky Mountains National Park." *Journal of Wildlife Management* 75, no. 5 (2011): 1145–54. www.jstor.org/stable/41418148.

## Chapter 15. Tundra Swans and Snow Geese

Bland, Sam. "A Wildlife Spectacle at Pungo Lake." CoastalReview.org, February 2, 2012. https://coastalreview.org/2012/02/a-wildlife-spectacle-at-pungo-lake/.

Carson, Rachel. "Mattamuskeet: A National Wildlife Refuge." *US Fish & Wildlife Publications* 5 (1947). https://digitalcommons.unl.edu/usfwspubs/5.

Coastal Wildlife Refuge Society. "Charles Kuralt Trail." Accessed February 14, 2024. http://npshistory.com/brochures/nwr/north-carolina-charles-kuralt-trail.pdf.

Cornell Lab of Ornithology. "Tundra Swan Overview." All About Birds. Accessed February 12, 2024. www.allaboutbirds.org/guide/Tundra_Swan/overview#.

Earnhardt, Tom. *Crossroads of the Natural World.* Chapel Hill: University of North Carolina Press, 2022.

Elder, Renée. "David and Goliath: The Fight to Site an Outlying Landing Field in Washington and Beaufort Counties." *North Carolina Insight*, February 2006, 106–0. North Carolina Center for Public Policy Research. https://ncppr.org/wp-content/uploads/2017/02/David_and_Goliath-The_Fight_to_Site_an_Outlying_Landing_Field....pdf.

Holt, Craig. "Navy Calls Truce, Ends OLF Battle." *Carolina Sportsman*. Accessed February 12, 2024. www.carolinasportsman.com/hunting/waterfowl-duck-hunting/navy-calls-truce-ends-olf-battle/.

Jeffers, Robinson. "Love the Wild Swan." In *Robinson Jeffers Selected Poems*, 59. New York: Knopf Doubleday, 1965.

Kinsella, Lindsey. "The Dwindling North Carolina Red Wolf Population." *US Environmental Policy* (blog), Nicholas School of the Environment, Duke University, April 2, 2020. https://blogs.nicholas.duke.edu/?s=kinsella.

Lawson, John. *A New Voyage to Carolina*. Chapel Hill: University of North Carolina Press, 1967, 150.

Satore, Joel. "Tundra Swan." Photo Ark, *National Geographic*. Accessed August 10, 2024. https://www.nationalgeographic.com/animals/birds/facts/tundra-swan.

Wilson, E. O. *Half Earth: Our Planet's Fight for Life*. New York: Liveright, 2016.